CHAOS AND COSMOS

HEIDI C.M. SCOTT

CHAOS AND COSMOS

Literary Roots of Modern Ecology in the British Nineteenth Century

The Pennsylvania State University Press
University Park, Pennsylvania

All poetry quotations and extracts come from the Poetry
Foundation website, http://www.poetryfoundation.org,
unless otherwise noted.

Library of Congress Cataloging-in-Publication Data
Scott, Heidi C. M. (Heidi Cathryn Molly), 1978- author.
Chaos and cosmos : literary roots of modern ecology
in the British nineteenth century /
Heidi C. M. Scott.
p.   cm
Summary: "Traces an epistemological legacy
from Romantic and Victorian ecological literature to
modern scientific ecology. Investigates two essential
and contrasting paradigms of nature that continue
to be debated today"—Provided by publisher.
Includes bibliographical references and index.
ISBN 978-0-271-06383-6 (cloth: alk. paper)
ISBN 978-0-271-06384-3 (pbk. : alk. paper)
1. English literature—19th century—History and criticism.
2. Nature in literature.   3. Ecology in literature.
4. Chaotic behavior in systems in literature.
5. Microcosm and macrocosm in literature.
6. Romanticism—Great Britain.   I. Title.

PR468.N3S38   2014
820.9´36—dc23
2014004459

The Pennsylvania State University Press is a member of the
Association of American University Presses.

It is the policy of The Pennsylvania State University Press
to use acid-free paper. Publications on uncoated stock
satisfy the minimum requirements of American National
Standard for Information Sciences—Permanence of Paper
for Printed Library Material, ANSI Z39.48–1992.

For Colleen, who lent an ear and dotted my i.

In the realm of evolutionary biology and ecology, ours is an unpredictable world and our place in it an accident of history; it is a place of many possibilities that are influenced by forces beyond our control and, in some cases at least, beyond our immediate comprehension.

—RICHARD LEAKEY AND ROGER LEWIN, *The Sixth Extinction* (1996), 230–31

Circumstances are like Clouds continually gathering and bursting. While we are laughing the seed of some trouble is put into the wide arable land of events. While we are laughing it sprouts, it grows and suddenly bears a poison fruit which we must pluck.

—JOHN KEATS TO GEORGE AND GEORGIANA KEATS,
14 FEBRUARY–4 MAY 1819

Contents

Acknowledgments

Many minds contributed to this book at its various stages from shadowy notion to monograph. I would like to thank my mentors and colleagues at the University of Maryland. Jeanne Fahnestock, Orrin Wang, and major advisor Neil Fraistat suggested primary texts, combed through drafts with patience and interest, and commented with insistence. Compatriots Kate Singer, Joseph Byrne, and Elizabeth Whitney helpfully interrogated my ideas in the earlier drafts. My colleagues at Florida International University, especially Meri-Jane Rochelson, Kathleen McCormack, and Yvette Puggish, provided insights about the introductory discussion that clarified the whole project. My parents, Tom and Bonnie Kime Scott, provided the same intellectual support they always have, with perceptive readings of the introduction and the Keats section. My wife, Colleen Flaherty, a biological scientist and environmental policy expert, helped me recognize and elaborate on the essential disciplinary crossings blazed in these pages. The peer reviewers' comments were crucial to the final organization of the book, and I thank them for their time and fair appraisals. Penn State University Press editor Kendra Boileau and her associates were a joy to work with in the last stages, as we honed the manuscript for its final audience, you, the reader. Thank you all.

Two Portraits of Nature

Ecological theories offer ideological portraits of nature. As portraiture can disclose an individual's origins, talents, deeds, powers, and disposition, not just his or her physical appearance, various portraits of nature can be fantastically contradictory. The classic ecological paradigm of nature favored well into the twentieth century depicts a character that is fundamentally balanced, nurturing, and intelligible, with a face that changes only gradually. More recently, postmodern ecology depicts nature as inherently chaotic, stochastic, and subject to catastrophic change: a character with an unstable personality.

Imagine these two distinct aesthetic cohorts, one temperate and one tempestuous, on display at an exhibit. The first wall holds representations of nature as an old-growth forest, a fanning coral reef, an entangled riverbank. Subject to simple natural laws, its delicate balance emerges from biodiversity, and its changes are gradual and tend to articulate favorable adaptations. The nature of the balance paradigm is readily degraded by human impacts, and our life sciences investigate ways to tinker with imbalance to restore an original ideal, or climax, state. Balanced nature is also the landscape of human stewardship and dominion, hearkening back to Eden.

The second wall presents portraits of an atoll drowned by a hurricane surge, the fluorescent cascade of a lava flow, a plume of oil and gas rising from a pipe on the sea floor. Predictable only through probabilistic calculus, changes are rapid, nonteleological, and subject to the impact of chance events. These portraits of chaotic nature are sensational and awe-inspiring. They partake of the aesthetics of the sublime by making the observer feel small and powerless. Ironically, these sublime forces of change that make humans feel small within the shadow of rioting nature are, in postmodern ecology, now projecting our own activities onto the canvas. *Homo sapiens* have gone from being the observers of sublime chaos in nature to being co-authors of it. "Natural disaster" and "act of God" will never have that simple sense of passive innocence they had before the Industrial Revolution and consequential climate change. Even though we know that climate is not the same as weather, a shift in consciousness in the twenty-first century makes every flood, drought, hurricane, extinction, famine, disease, and invasive species potentially traceable to our balance sheet. The paradigms of classic balanced nature and postmodern chaotic nature are two ways of portraying an immensely diverse and complex natural world. Ecological paradigms are never strictly objective; they are colored by the cultural conditions of their emergence.

Within the past half century, ecological science has critiqued the balance paradigm as a misleading, quasi-mystical construct that forces economic and mechanical models on the obscure dynamics of ecological interconnection. A major trend in ecology today identifies at least three important departures from this classic paradigm, in effect navigating between models of balance and cataclysm. These ideas are not universally accepted by ecologists, but they have become mainstream theories of natural patterns.[1]

The first finding is that ecological communities are shaped by chaotic and random forces. Population dynamics and species distribution must be understood through stochastic processes that make changes in environments difficult to predict. Landscapes are thought to be composed predominantly of species mosaics wrought by chaos and chance rather than communities united by synergy and mutualism.

Second, evolution is not only based on adaptation but, at least as significantly, reacts to environmental contingency. Extinctions and rapid shifts in

---

1. Two valuable book-length studies by ecologists give further detail to this distinction between the old balance paradigm and the new chaotic model of change in nature: Botkin's *Discordant Harmonies* and Kricher's *Balance of Nature*. Leakey and Lewin's *Sixth Extinction* contains an informative chapter on stability and chaos in nature. For a review of evolutionary patterns in theory, see Gould and Eldredge.

morphology can be explained more effectively through a rubric of catastrophe and random drift than through the gradual and intelligible articulation of superior adaptations over deep time. Extinction tends not to be the result of interspecies competition, with the superior form winning out, as Darwin claimed. Instead, most extinction is due to random environmental disturbances on many scales, from regional ripples to global cataclysms. As a consequence, adaptation is itself radically contingent upon circumstances, and most evolution takes place only after disturbance. This theory of evolution by punctuated equilibrium is a revision of Darwinian gradualism in the era of chaos ecology.

Finally, human impacts are the most important factor in ecological disturbance as we move into the twenty-first century. The interscalar impacts in our era include greenhouse gas emissions, habitat appropriation, deforestation, chemical changes in oceans and fresh water, intense harvesting of fisheries, and industrial agriculture driven by petrochemicals. Ecology's most pressing questions come from the exigencies of human impacts on the biosphere; these impacts are superadded to the scientific proposal that the background character of nature is itself chaotic. Chaos ecology revolutionizes classical views of a balanced natural world that have dictated scientific perceptions since at least the Enlightenment. For better and worse, it is the new creative principle in ecology, and its roots reach down into Romantic-era soil. This study explores literary expressions of ecological chaos starting in the Romantic period. It claims that nineteenth-century literary narratives played a seminal role in sketching out the postmodern view of chaotic nature that would emerge in ecological science of the late twentieth century.

The microcosm, another scientific concept with Romantic precedents, works in this study as the counterpoint to chaos ecology. Microcosm is an empiricist's tool for modeling ecological processes that often relies upon the ideal of a balanced nature. The physical and theoretical constructs available in microcosms helped ecology become an experimental science in the twentieth century, moving past the methods, based upon observation and cataloguing, of earlier natural historians. Microcosm experiments propose that ecologists can build, maintain, and manipulate small systems in order to shed light on the complex dynamics of nature in larger, real-life scales. Ecological microcosms are domesticated, simplified ecosystems: mechanistic models that serve as proxies for natural environments. They are particularly useful in the study of disturbance because they can be used to model pollution, extinction, and other stress gradients that ecologists would not want or be able to manipulate in real environments.

Still, microcosms are ideal constructs that assume that there are discrete environments in nature, such as ecosystems with closed communities, rather than a continuum of greater and lesser similarity of form that ranges across the globe. The circumscription of discrete small worlds is itself a conceptual convenience that is imperfectly reflected in the biosphere. By experimenting with species composition and with chemical and energetic balances in the model, ecologists hope to discover what causes underlie disturbance and degradation in larger environments that had once been stable. These two figures of thought, chaos and the microcosm, have a theoretical role in the debate over nature's character as chaotic, balanced, or some combination of the two, as well as an applied basis in the actual methods of experimental ecology.

I propose that the seed of imagination that would enable a scientist to study a lake as a microcosm at the formal, empirical level was sown by poets of the nineteenth century who consciously drew a sphere around small-scale nature in order to make sense of spots of time and place amid the increasingly chaotic, global, industrial modern world. This book interrogates the literary origins of the two tropes, and how they have been transcribed into the sciences of nature. It proposes that innovative nineteenth-century narratives of ecological disturbance foresaw chaos ecology at a time when gradualism and balance were paradigms of natural history. It also proposes that nineteenth-century poets helped scientists conceive of ways to simplify nature in microcosm without dismembering its complex structures. Scientific reductionism tends toward atomies and dismantled systems, but the microcosm attempts to reduce organic systems without dissecting them. Microcosm experiments are akin to a particular kind of poetic lyrical holism, in both image and prosody. These two tropes, the chaotic narrative and the microcosm model, effectively align disparate portraits of nature in nineteenth-century literature, and bring the nascent ecological sciences into a dialogue with literary prophecy. They make ecological theory interdisciplinary in the distinct arenas of narrative (natural history) and the structured complex poetic image (empirical design).

From this study of literature and ecological science, we may draw the conclusion that the imagination at play in literature provides an alternative, perhaps richer, form of modeling ecological change, especially the nonlinear change seen in chaos ecology. Although the various forms of scientific modeling are essential to predicting how impacted systems will change, the constraints of artificial and reductive design can also cause problems when the models stand in for natural systems and become the major focus of the science. Perhaps literary microcosms represent a more organic form of conceiving ecological systems that is valuable to theories of natural dynamics currently dominated by the scientific model. Because literary models are dedicated neither to reducing complex systems to their barest minimum, nor to abstracting them to accommodate millions of variables, they may remain expansive and fluid, they may preserve historical memories of ecosystems, and, by their articulate poetic form, they may compel readers to attend, comprehend, and care for the actual natural space they contain in prosody and imagery.

Chaos and Cosmos raises the stakes on ecocriticism's claim that literature generates essential knowledge about nature complementary to our scientific views. I contend that literature in the early decades of industrialism achieved a unique narrative perspective on the transformation of landscapes, and that poets began to model natural systems as empirical entities contained within the natural parameters of prosody. This perspective is continuous with scientific ecological methods generated over the course of the twentieth century. Ecocritical perspectives often oppose literary thought (and its catalysts, inspiration and imagination) to scientific method (design and repetition); this outworn creed ignores the affinities of investigation based on intimate knowledge of ecosystems. Both writers and ecologists are close readers of natural systems, and both use imagination to rework cryptic natural processes into coherent theories that elucidate patterns—even chaotic patterns.

Moreover, the British nineteenth century provided a unique nexus of cultural, historical, and disciplinary crossings that allow us to look back on a cohort of writers not only as poets sympathetic to natural forms, but as investigators of a changing landscape. For example, Dorothy Wordsworth is at once a poet, diarist, natural historian, and social ecologist during a time of war and revolution. In addition to being an empiricist, Charles Darwin is a storyteller who crafted the most important narrative of the nineteenth century out of a wealth of disparate case studies. Richard Jefferies's immersive writing on the rural nature of Wiltshire was colored by his mid-nineteenth-century context, which imposed industrial transformation and the despoilment of the British countryside on his otherwise idyllic close readings.

Environmentalist ideals are often woven in with ecological paradigms, but this book is focused on the ways in which we know nature (ecological epistemology) more than on the ways in which we ought to act within it (environmental ethics). Of course, literature of the environment is laced with ethical convictions, and scientific ecology inevitably is, too, so this division between epistemology and ethics is permeable. Ecologically minded people are not only scientists but are also nature writers, dumpster divers, environmental justice advocates, urban gardeners, and annotators of almanacs. We are people living within acculturated nature—which includes wilderness areas, farms, suburbs, exurbs, and cities. Nonetheless, there is an important distinction between "ecologist"—one who studies the interactions between organisms and their environment—and "environmentalist," one who advocates an ethics-based set of human behaviors within nature. *Chaos and Cosmos* circles around literary and scientific ecologists.

Ecocriticism often wrangles with its own limits of authority, questioning whether literary critics are in a position to comment on scientific epistemology outside the conventional bounds of metaphor, trope, rhetoric, theory, and history. Greg Garrard's influential primer on the field recognizes ecocriticism's close relationship with the science of ecology, but his stance is one of subordination when it comes to elucidating "problems in ecology": "Ecocritics remain suspicious of the idea of science as wholly objective and value-free, but they are in the unusual position as cultural critics of having to defer, in the last analysis, to a scientific understanding of the world" (10). Certainly, the general point Garrard is making is valid: ecocritics are in the tricky position of supporting scientific findings (like climate change) while maintaining a strategic distance that allows for critical perspectives on inherent subjectivity and gender, sex, race, and species biases in the practice of science. Literary studies, even those based on ecology, have a reputation for being antiscientific: a dangerous rap that we must openly disavow. One of the challenges to ecocriticism that Garrard outlines is that the field needs to "develop constructive relations between the green humanities and the environmental sciences" (178). In particular, ecocritics need to address the inconsistency between literary pastoral or Gaia-inspired views of nature in harmonious balance and postmodern ecology's view of nature as inherently dynamic and unpredictable (178). I propose that ecocriticism can do better than play the role of duplicitous sibling to ecological science. Ecocriticism can theorize how the scientific understanding of nature has literary origins. Literature begat methods of narrative and modeling in ecological science via seamy collaborations with philosophy, natural history, and the established natural sciences of physics, chemistry, and biology. When this interdisciplinary argument is accepted, literature is promoted to progenitor of our scientific understanding of the natural world. As an ancestor, literature shares responsibility for the very biases that humanities scholars hasten to expose in their analyses of science.

In recent years ecocritics have followed many intriguing links between nature and literature that go beyond the classic vision of first, or primary, Nature—the austere wilderness ideal of nature "out there," which Kate Soper has also called "metaphysical nature" (155–56). Replacing the mythos of the wilderness is the contemporary vision of second nature—an environment entangled with human uses, which in our time involves disturbance, degradation, and chaotic change. Primary Nature is a proper noun—a construct of the entirely extrahuman. It is a wilderness that no longer exists. Second nature is the set of environments that we actually dwell in, cultivate, enrich, and despoil—what Soper has dubbed "lay nature." Between metaphysical Nature and lay nature is the material system that is the natural sciences' object of study, what Soper calls "realist nature." Literary ecocritics in particular should be interested in claiming this realist nature as an object of investigation for literature as well as the sciences.

Ecocritics cut their teeth on metaphysical Nature in the latter part of the twentieth century, with special concentration on German, British, and American Romanticism. In the past decade or so, lay nature has succeeded as the most important locus for the attention of ecocriticism, especially as the field has turned toward lived-in environments built around postcolonial, socioeconomic, queer, and industrial-era revisions of pure Nature. Even ecocritics with British Romantic concentrations, such as Jonathan Bate, Alan Bewell, Timothy Morton, and James McKusick, have in recent years written extensively on disturbed environments rather than pristine ones. Bate's *Song of the Earth* is the shiniest green among the four, but his book includes an important reading of disturbance surrounding the 1815 Tambora eruption extending to Keats in 1819 (104–5). Bewell's book on colonial disease transmission has important implications for the spread of ecological calamity in a global economy. Morton's adoption of "dark ecology" expounds on the obsolescence of Gaia and harmony in preference for a mournful intimacy with ecological sickness. The work of these four ecocritics (as well as many others working in American and non-Western literatures) highlights the kinship between environmental literature and close reading, historical concurrence, and literary theory. James McKusick's *Green Writing* is an important contribution to transatlantic Romantic theory that details the ways in which American nature writers inherited the ideas of their British forerunners. McKusick comes the closest to my interest in the dual characterization of nature as balanced and chaotic, and, like Jonathan Bate, he occasionally uses scientific discourse to elucidate literary texts. For example, he asks the reader to imagine Romantic literary society as an eco-system, "a vibrant community in which competition and synergy, exchange of ideas and flow of information, predators and prey, hosts and parasites, all coexist in the turbulent vortex of a shared environment" (18). The analogy of societies as ecosystems seeking synergy and troubled by chaos is a mainstay of human ecology. However, these literary studies do not pursue a specific claim about the literary origins of ecological science. As a fully interdisciplinary study, *Chaos and Cosmos* pays more attention to ecological studies than is conventional in literary criticism, and conducts more readings of novels and poems than any scientific work would do. It traces the ancestry of ecological science to find lurking literary forebears.

Poems and novels can elucidate the material processes, species relationships, and tempo of change ongoing in the physical world. We generally expect science to conduct this investigative work. Instead, I would argue that literature provides insightful systemic readings of physical nature that often predate scientific attention. The aim of this book is to establish how literature was involved in formal explorations of realist nature before scientific ecology existed. The nineteenth century of industrialism and colonialism undid the capitalization of Nature as an austere proper noun. This semantic change to "nature" demonstrates how some literature of this period challenged the classical paradigm of economic balance before ecological science had its methods in place. These two essential literary tropes, chaos and the microcosm, have evolved over the past two centuries into theories and methods in ecology.

Around 1887, near the end of his short life, the British writer and naturalist Richard Jefferies penned a precocious observation on the tension between the paradigms of balance and chaos, which he called "The Absence of Design in Nature":

> When at last I had disabused my mind of the enormous imposture of
> a design, an object, and an end, a purpose or a system, I began to see
> dimly how much more grandeur, beauty, and hope there is in a divine

chaos—not chaos in the sense of disorder or confusion but simply the absence of order—than there is in a universe made by pattern. This draught-board universe my mind had laid out: this machine-made world and piece of mechanism; what a petty, despicable, microcosmos I had substituted for the reality. Logically, that which has a design or a purpose has a limit. The very idea of a design or a purpose has since grown repulsive to me, on account of its littleness. I do not venture, for a moment, even to attempt to supply a reason to take the place of the exploded plan. I simply deliberately deny, or, rather, I have now advanced to that stage that to my own mind even the admission of the subject to discussion is impossible. I look at the sunshine and feel that there is no contracted order: there is divine chaos, and, in it, limitless hope and possibilities. (*Old House at Coate* 163)

This passage, vehement and celebratory, lays out the organizing principle of the present study. Jefferies's divine chaos recovers hope from Victorian angst by substituting the sublime splendor of infinite creativity for a preordained mechanistic cosmos. To be designed or purposeful, as he calls the "microcosmos," is to be static, inorganic, regulated. Critiquing at once the religious conviction of divine Providence and the Enlightenment predilection to see nature as a grand machine, Jefferies asserts that the "absence of order" is a larger, liberating view of an organic natural world. Machines for industrial tasks are what humans sketch out on their drawing boards, but, by analogy, to reduce the earth to a "piece of mechanism" is to leech away the lifeblood of the vital, chaotic cosmos. Microcosms that model ecological processes occur in both literature and science. They serve to reduce the complexity of open natural systems to simplified, intelligible model systems. What is often sacrificed is the creativity, the serendipity, the breaking down of borders and limits enabled by the paradigm of a chance-driven and design-free nature.

In art theory, randomness has taken on positive connotations of serendipity, complexity, and unscripted authenticity. Akiko Busch describes the serendipity of craft, where artists cannot totally control the chaos of the wheel and the glaze colors that emerge from the kiln, and woodworkers seek out the unique grains and shapes that weather and climate impose on their medium. The reconciliation of randomness, of chaos, with design and control is an essential source of artistic creativity (75). Environmental artist Andy Goldsworthy crafts his pieces within the happenstance conditions of open settings, so that unpredictably changing winds, stream flows, light, and temperature play an essential role in the formation and dissolution of his work; he welcomes the chanciness of art al fresco. The poet Gary Snyder has written on the chaotic reciprocity between nature and language. Complexity in evolved wild systems, Snyder writes, "eludes the descriptive attempts of the rational mind. 'Wild' alludes to a process of self-organization that generates systems and organisms, all of which are within the constraints of—and constitute components of—larger systems that are again wild, such as major ecosystems or the water cycle in the biosphere. Wildness can be said to be the essential nature of nature. [. . .] So language does not impose order on a chaotic universe, but reflects its own wildness back" (174). Jefferies, Busch, Goldsworthy, and Snyder all celebrate randomness for its capacity to rupture the comfortable quotidian, one of art's signal intents. They prefer portraits of nature in chaotic dress, where our human dominion within the elements may at any moment be challenged or overthrown, and where the pastoral idyll falls away to reveal a creative unknown. Postmodern nature introduces art and design theory to the chaotic muse. It is the radical denouncement of Ecclesiastes 1:9: "That which has been is that which will be, and that which has been done is that which will be done. So there is nothing new under the sun."

Resolving Opposition

In the spirit of disciplinary unity, this book is about the literature of ecological reconciliation. Chaos and the microcosm are complementary figures of thought that help us understand the dynamics of our disheveled home, or *oikos*. One is a temporal narrative of chaotic change; the other is a spatial model of balanced exchange. At a surface level, the two tropes appear as aesthetic complements whose relationship is based on this essential contrast. To a certain extent, they are just that. The microcosm contains; chaos overflows. Microcosms are Quaker hymns of self-sufficient simplicity; chaotic systems conduct matter and energy in the mode of postmodern symphonies. Microcosms are domesticated pets; chaos is a beast in a looming shadow. But if we plunge a little deeper into the conceptual pool, we find strange currents that confuse and conjoin these tropes. Microcosms in ecology, as simple and closed systems, are always susceptible to major shifts if certain players gain greater dominion. By virtue of their diminutive size and simple composition, they often lack the chemical and biological buffers that tend to keep systems stable through small fluctuations. A shallow lake, the classic microcosm in nature, can shift from pellucid clarity to a plankton-choked morass if the water receives just a bit more sun or nitrogen. An aquarium will be overcome by algae when its detritus-eating snail dies. Delicate balances, while provisionally self-sustaining, are perilously close to dissolution; both balance and rapid degeneration are vying fates in microcosmic systems. While we may not find aesthetic pleasure in the slimy aquarium or the weedy pool, an ecologist can show how this microcosm has spontaneously evolved into an alternative stable state, where a new clutch of species controls the system. Ecological microcosms are subject to chaotic fluctuation.

Chaotic dynamics, by contrast, connote higher organization and eventual coherence based on minute, unpredictable variations in an initial system. This spontaneous new order emerges from newly discovered affinities among components and the power of initial conditions to organize the emergent structure at higher levels. The study of chaos need not be cloistered in the esoteric symbol languages of mathematics. Chaos theorist Ilya Prigogine enriches history by describing how technological innovations such as the advent of steamships in the nineteenth century can create their own niches in the ecology of economics. Innovations that provide major practical advantages can "transform the environment in which they appear, and as they spread, they create the conditions necessary for their own multiplication, their 'niche'" (Prigogine and Stegners 196). Chaotic modeling can demonstrate how patterns of urbanization and rural depopulation are directed by positive feedback and nonlinear dynamics; the city grows out of the general store where two roads once happened to cross. Chance factors break the bland symmetry of population distribution based purely on diffusion, but chaos replenishes pattern by providing strange attractors at the crossroads. Small historical events like when and where the steam engine was invented (1712, England) have the power to revolutionize global society along a new set of parameters in only three centuries. Urban settlement, industrial work time, fossil fuel consumption, new class dynamics, population booms, mass transportation, the modern economic imperative of expansion, and most of the ecological disturbances we face today are downstream of this historical happenstance. One small trickle of technology found favorable conditions and nudged into motion this major ontological shift from the long established environments of human evolution to this strange state of global industrialism.

With any progressive history, there is a danger of retrospective bias awarding destiny to a "chosen" or "superior" culture. Such is the bent of most heroic histories of imperial Britain, and of some of the more shameful interpretations of evolutionary theory. But this false telos involves inadequate factoring of the chancy initial conditions, what John Keats saw as the vanguard of circumstance into which a seed of future events happens to be sown. A less celebratory narrative of chaotic dynamics emerges from the epidemiology of measles and cholera in London, another offspring of industrial and colonial dominion.

In these literary readings of ecological chaos, there is no intention to claim that they achieve formal mathematical chaos, discovered in the 1960s. The vogue of chaos theory as a new way to read patterns in many disciplines—from the fine arts, to literature, to law—has caused some grumpiness among mathematicians who would like to sequester chaos theory within their discipline. More loosely interpreted, chaos as a metaphor allows for a third vision of emergent structure that was invisible in the classic dichotomy between order and disorder. It shows the limits of human control over natural systems. Michael Crichton's character Ian Malcolm, a mathematician, repeatedly brings chaos theory into his perceptions of dinosaur behavior, island ecology, and the inevitable failure of Jurassic Park's security. In the 1990s, Katherine Hayles demonstrated how the trope of chaos can be applied to twentieth-century narrative theory. Hayles is careful to maintain that balance makes an idea like chaos particularly useful: chaos does not obviate order; it merely reorients our understanding of how order occurs in the natural world by signaling the prevalence of slight variations. She invokes images of environmental problems brought into focus by the trope of chaos: "Industrial pollutants are released into the atmosphere; along with carbon dioxide, also a by-product of technology, they create the greenhouse effect; the resulting climate changes wreak havoc with the global ecosystem. Cascading effects from initially small causes could, and have, been observed at any time. But whereas in earlier epochs they tended to be seen as anomalous or unusual, now they are recognized as paradigmatic of complex behavior" (15).

This metaphor of chaos can also be used to show how writers began refiguring nature during the industrial shifts of the nineteenth century. There are intriguing moments when writers anticipate ecological concepts that have since been formalized under mathematical chaos theory, such as population dynamics and meteorology, but in general the chaos trope indicates an author's vision of the natural world that falls poignantly in an alternative state between perfect order and utter randomness. Recall Richard Jefferies celebrating his break from the limits of the mechanical microcosmic worldview to embrace the more radical possibilities of dynamic chaos.

This metaphor of chaos reconciles the paradigms of nature, balance versus disorder, thereby providing a conceptual frame for the work of adventurous nineteenth-century writers who diverged from long-standing cultural assumptions about the economy of nature, static created species, and landscapes impervious to human activity. A few exceptional Enlightenment theories of a dynamic natural world aided in nineteenth-century literary visions of chaos. Geologists like Georges Cuvier advocated catastrophism over gradualism; volcanologists, including Humphry Davy and James Smithson, argued that the atmospheric impact of eruptions mimicked industrial emissions; and natural historians like Alexander von Humboldt began to formally observe the impact of climate and altitude on ecological interrelations.

The narrative chaos at play in this book refers to the ecology of disturbance: how fiction envisioned disturbed nature emerging downstream in time. The British nineteenth century is particularly rich with second natures, those landscapes that became scattered and smothered by the cultural productions of industrialism. Wrecked environments still have ecologies, and literary ecocriticism is actively involved in theorizing how these second natures have become the focal environments of concern in the twenty-first century. For example, protest lyrics like Gerard Manley Hopkins's "Binsey Poplars" highlight the pathos of landscapes sacrificed to economic advance, when the trees that had been planted by human hand to give river shade are "All felled, felled, are all felled / [. . .] Not spared, not one / [. . .] On meadow and river and wind-wandering weed-winding bank / [. . .] After-comers cannot guess the beauty been" (lines 3, 5, 8, 19). The Darwinian entangled bank is exposed and degraded into a muddy riverside slide by the elimination of the keystone trees. In this poem, ecological complexity is hacked down into a simpler utilitarian economic order that gives easy access to river transportation. Chaos is in the moment of the disorderly deed—the felling of the Binsey poplars—as well as in the implied narrative of its long-term consequences: increased storm runoff, erosion of the banks and nearby soils, decreased habitat for spawning fishes and nurseries, accelerated river flow, and, most apparently to Hopkins, the loss of the beauty of the "sweet especial rural scene" and its emotional gravity (line 24). That chaotic moment of the felling will determine the character of the landscape and the mentality of its human inhabitants for generations to come.

There is a microcosm effect, too: this minor act of landscape disturbance models the larger process at work, changing the English countryside into a more economically productive environment. The hedgerows, for example, were actively decimated throughout the country following World War II to accommodate the needs of modern agriculture. (An enormous combine machine cannot operate in the same small space as a plowman and his horses.) A student of ecology could use Hopkins's before-and-after as a microcosm study of the effects of riparian deforestation. The minor anecdotal poem becomes a vision of the future beyond the lyrical moment in 1879, a narrative of a new landscape downstream from disturbance. Note that it need not be Nature, wilderness, at the starting point of degradation. Even second natures, like the row of poplars planted by humans long before Hopkins's time, develop into fixtures in the landscape—ecologically and emotionally.

The present project seeks to recover two highly formalized scientific tropes from their twenty-first-century denotations and restore them to their original homes in interdisciplinary philosophy. Therefore, my use of the terms *chaos* and *microcosm* throughout the analysis gestures to the evolution of a set of connotations associated with these concepts over the course of the nineteenth century. From mythological chaos, the epitome of vile incoherence, arose the intriguing paradox of higher levels of order; one of these structures is biological life itself. Analogously, the concept of an isolated, coherent ecological microcosm that could model larger dynamics arose from the ancient philosophical construct of human bodies as little words resembling the larger cosmos. The poems and other texts selected for this study each demonstrate a specific way in which a returning trope can serve to organize the thoughts of a culture struggling with new phenomena. Literary tropes are returning motifs that not only aid in the communication of ideas but have also frequently been identified as constitutive of experience (Ortony 253). Conceptually crucial tropes of the imagination inform the development of inchoate sciences lacking foundations in theory, such as the ecology of the nineteenth century.

I stop short of claiming that a direct causal relationship exists between the evolution of these literary tropes in the nineteenth century and their subsequent adoption into scientific epistemology. However, I maintain that British culture, in order to develop a discourse around the natural world newly altered by industry, had first to create theoretical scenarios and frames of reference using the literary imaginary. These chaotic narratives and microcosmic models became tools shared by an intellectual culture reacting to the environment as it changed under industrialism. The science of ecology is only the most recent method we have developed to examine nature, and it has inherited methods from many benefactors.

PART 1

Chaos

1

Four Histories of Change

The following two chapters read four British narratives of ecological disturbance—two Romantic and two Victorian. All four works show an innovative use of catastrophe to describe the state of the natural world and its
tempo over time. In contrast to the paradigm of balance in nature, these
four works experiment with the power and potential inventiveness inherent
in a chaotic worldview. Chaos is the loss of control, but it is also the discovery of a new kind of infinity.

These works were inspired by specific moments in history when natural
disasters stressed human conceptions of nature in balance. They form a
series of footsteps across the long nineteenth century, each taking a sample
of the current cultural and natural climate. Gilbert White's *Natural History
and Antiquities of Selborne* dates to the early Romantic 1780s; Mary Shelley's
The Last Man emerged in the late Romantic 1820s; Richard Jefferies's *After
London* is a work of the Victorian 1880s; and H. G. Wells's *The Time Machine*
straddles the Victorian and modern periods at the turn of the twentieth century. These selections represent important crossroads between natural and
human history because they were written in the wake of actual cataclysms.
Three of the four works follow immense volcanic eruptions. Laki (1783),
Tambora (1815), and Krakatoa (1883) had some of the most devastating global
effects ever seen from eruptions, with tens of thousands of people killed in

each event. Some estimates put the human death toll in the years following the eruptions in the hundreds of thousands (Walker, citing Grattan and Brayshay). Each eruption had acute and chronic effects across the globe, including anomalous cold weather, acid rain, crop failures and famine, and the spectacular optical effects of the ash suspended in the troposphere. Though balance, gradualism, and coherent evolution were popular constructs in British nineteenth-century culture, these authors wrote about chaos while under the influence of natural disasters. Their immediacy is particularly relevant to our ecological outlook in the twenty-first century with climate change. Gilbert White in the 1780s assumes that he will find a static, perennial nature in Selborne and instead finds surprising lurches in animal populations, extreme temperatures, and the Laki eruption flaring over the parish. Mary Shelley's fiction imagines an apocalypse of the plague as globalization spreads disease transmitted via trade routes. The sickness is incubated by a perversely pleasant warm climate, but the lingering psychological effect of Mount Tambora's eruption, which clearly marked *Frankenstein*, is still suspended in the atmosphere of *The Last Man*.

In the next chapter, we will explore Richard Jefferies's invention of a postapocalyptic pristine nature, where ecological succession has engulfed the relics of Victorian civilization but the toxic legacy of industry skulks just below the surface. Jefferies was witness to British food shortages from crop failures due to Krakatoa's volcanic winters in the 1880s, and he was openly intrigued by environmental chaos's power to reinvent a degraded world. H. G. Wells did not write under the influence of an eruption, but he uses the theory of evolution by natural selection to imagine a harrowing connection between Victorian industry and an actual machine nature within a million years. In this story, the initial conditions of industrial economic inequality have chaotic downstream effects on human evolution.

I have divided the four works into two separate chapters on chaos because the traditional transition from the Romantic to the Victorian period around midcentury also signals an upward shift in the intensity of industrial activity and its ecological effects in England. In the earlier writing of White and Shelley, industry is a mere shadow beyond the borders of the narrative. By the second half of the century, Jefferies and Wells designate industrial pollution and machinery as central elements in future nature. A moral thread that distrusts the influence of industrialism colors their writing, and this thread is absent in the Romantic works. Jefferies and Wells also had Darwin's ideas at their disposal, so they are necessarily in dialogue with natural selection. Before delving into the Victorian novels, I discuss how contemporary atmospheric science was beginning to make connections between volcanic eruptions and factory emissions as comparable forms of air pollution.

In all four works, conceptual innovations of chaotic nature foreshadow theories in modern ecology, including population ecology, succession dynamics, disturbance mosaics, and climate change. Purposely omitted in this study are cozy catastrophes such as Samuel Butler's *Erewhon* (1872) and William Morris's *News from Nowhere* (1890). Although these works show a Victorian disaffection for urban industrial settings, their simplistic "back-to-the-earth" dreamworlds are not particularly enlightening in relation to ecological studies in the twenty-first century. Their utopianism is classical rather than modern.

The literary use of catastrophe patterns requires a move beyond the calm quotidian of drawing room and garden settings. British novels of the nineteenth century mostly conformed to the convention of nature's constancy beneath human historical turmoil, from Scott to Austen, Eliot to Gaskell. Nature may change cosmetically from industrial blackening, as in Gaskell's *North and South* (1855), but polluted nature is not the central concern; character self-realization is. Where classic novels like *Bleak House* (1852–53) and *Middlemarch* (1874) prize coherent conclusions that advance moral arguments centered on human action, Romantic and Victorian literary scholarship has said less about contemporaneous novels with a modern vision of nature. *The Last Man* (1826) and *After London* (1885) are perhaps inferior works of literature when character and plot are the major considerations, but in exchange they propose that nature has become its own tragic character in a new age of ecological stressors: nature is dynamic, damaged, unpredictable, vengeful, enduring, and nurturing. The apocalyptic thriller can also afford a sense of triumph to readers. When not upstaged by nature, characters often show latent capabilities that come to shine in the new environment, as the old, outworn culture is turned under. Catastrophe provides an outlet for a different kind of romance, with fresh adversities and heroes showing adaptive spots and stripes: inoculated wanderers, new frontiersmen, and time travelers. The ecological thriller is receptive to subversive comedy and romance; it is not always a desperate slog through dystopian lands of extinction. Readers share the hope that catastrophe engenders unforeseen bounties, and that we are ennobled and strengthened by enduring its hardships.

In *The Order of Things*, Michel Foucault explores how a clearer under-standing of deep time enabled by new findings in geology and biology caused a rupture in the Western episteme of the nineteenth century, which had been built upon a long-standing assumption of natural history as a human-centered, providential narrative:

> [I]t was discovered that there existed a historicity proper to nature; [. . .] man found himself dispossessed of what constituted the most mani-fest contents of his history: nature no longer speaks to him of the cre-ation or the end of the world, of his dependence or his approaching judgment; it no longer speaks of anything but a natural time; its wealth no longer indicates to him the antiquity or the immanent return of a Golden Age; it speaks only of conditions of production being modified in the course of history. [. . .] The human being no longer has any history: or rather, since he speaks, works, and lives, he finds himself interwoven in his own being with histories that are nei-ther subordinate to him nor homogeneous with him. By the fragmen-tation of the space over which Classical knowledge extended in its continuity, by the folding over of each separated domain upon its own development, the man who appears at the beginning of the nineteenth century is "dehistoricized." (367–69)

In effect, the increasingly sophisticated life sciences were proposing a new and indeterminate paradigm of deep time in which human history was only one of myriad narratives. The human story was recent, heroic only from an egocentric point of view, and had neither clear origins nor a telos. The human story meandered, like all other life histories, through a pathless wood deprived of the landmarks that heroic history and religion had pro-vided. Foucault's perspective is postmodern and retrospective, and his read-ing pays little attention to how Darwin crafted a conscious narrative of purpose, articulation, and improvement to characterize evolution. Still, the kind of vertigo expressed by Matthew Arnold, "Wandering between two worlds, one dead, / The other powerless to be born," captures the more shadowy zeitgeist of Darwin's time, which Foucault identifies through his theory of the "dehistoricized" culture of the nineteenth century ("Stanzas from the Grande Chartreuse" lines 85–86).

The main thesis of these two opening chapters is that chaos in literature predates the scientific study of chaos in ecology. Eighteenth-century geologists, including Cuvier, Buffon, Lamarck, and (a little later) Lyell, had discovered both deep time and cataclysm in fossil evidence, but their discoveries were slow to be adopted into theories of the ongoing state of nature. From an ecological standpoint, these four literary works are precocious because they appraise catastrophic events in natural history and weave them into the fabric of futurity. It would overstress the interdisciplinary project to claim that these works demonstrate formal mathematical chaos. However, the very pattern of sudden ecological punctuation is a timely contribution of a literary imagination, as is the ecology of disturbance. These works are precursors to contemporary chaos ecology because they all insist that natural and anthropogenic disturbance must figure in our understanding of modern nature.

The Natural History and Antiquities of Selborne

Between 1768 and 1787, Gilbert White brought to the Enlightenment the first in-depth, in situ study of an ecosystem. The text has never been out of print since it was first published in 1789. White's *Illustrated Natural History and Antiquities of Selborne* reports on several aspects of Selborne's environment: its geology, botany, zoology, and climate. Notably for a man of the cloth, the Reverend Gilbert White never invokes God or any higher power in his attempts to explain mysterious forces of nature. He keeps his letters literal, detailed, and secularly speculative. White's narrative momentum is maintained through the progression of time, since the three dimensions of space remain fixed in his home parish. The single location distinguishes White's epistemological strategy from that of his contemporary Alexander von Humboldt, who traveled extensively to understand geographical relations in ecology. Observations while traveling gave Humboldt the theoretical grounds for biogeography, as he detailed the importance of elevation in the distribution of plant types in the Andes range. In contrast, Gilbert White's work is an early microcosm study because of his decades-long dedication to a circumscribed microenvironment. In a time when natural historians were engaged in a mania of exotic collection to fill cabinets of curiosities, White's enduring absorption in his home parish shows an impressive degree of concentration on the biodiversity in his immediate purview. The wanderlust of the colonial scientist did not influence White's own methods, though he kept up a

considerable correspondence with traveling collectors and scientific societies in London (Worster 6).

White's original use of phenology, the study of naturally recurring cycles
such as the seasons, provisionally advanced knowledge according to Enlightenment expectations of stability. A devotion to ornithology predisposed
him to detailing species migration according to predictable annual patterns.
In spite of its phenological design, the *Natural History and Antiquities of Selborne* deconstructs the Enlightenment sensibility of coherent, patterned
nature. By the end of White's chronicle, the author views extraordinary
events of ecological disturbance as essential to Selborne's natural state.
Critics have nearly always distilled White's text down to precisely the
inverse of chaos by celebrating its stable Enlightenment essence. The prevailing consensus on White's work is that his letters from Selborne articulate an Edenic vision of man living in contemplative symbiosis with his
natural surroundings. This balanced, preindustrial microcosm frame of reference massages the reader's shoulders with visions of simpler times to
which we may retreat, if only psychologically (Allen 50–51). Another perspective credits White with appreciating changes in nature, but only within
the comfortable boundaries of a taxonomic challenge, citing "his fascination
and delight in an ever-living yet ever-changing, ever-elusive, ever-miscellaneous
nature" (Bellanca 77).

There is more than fascination and delight in Gilbert White's letters:
readers also receive a ration of confusion, awe, and horror. Although the
first two-thirds of White's chronicle are passably at peace with the world
and imply the utopia of a stable and dynamic cosmos, to pin the whole work
within this frame of balance deprives White of the credit he deserves for
contemplating chaotic disturbance, the less comfortable mode of ecological
thought. Before the end of his quarter-century of correspondence, White
has grown into a more radical speculator on the complex dynamics around
him. If *Selborne* were really a chronicle recording eternal peace, it would be
functionally obsolete; a twenty-first-century visitor to the parish would recognize very little from White's account. *Selborne* is a classic text for modern
times not because it reinforces a set of established conventions about the
balance of Mother Nature, but because White successfully divests the balance paradigm in favor of a more modern view of nature based on discord
and contingency. The microcosm of Selborne, White discovers, was vulnerable to violent change and rapid degradation partially by virtue of its dimin-

utive scope. These theories of chaotic endangerment have not been developed in the critical literature on White's work.

By 1770, two years into his study, White is detailing the elaboration of precise methods of observation, and has coined for himself the title of "monographer" writing on a single scientific subject: "Men that undertake only one district are much more likely to advance natural knowledge than those that grasp at more than they can possibly be acquainted with: every kingdom, every province, should have its own monographer" (125).[1] White is wary of being overwhelmed by too much natural information and losing the clarity of an expert's vision in a specific sphere of knowledge. The Enlightenment filled shelves with surface-level, encyclopedic studies. Selborne has borders, and ample natural provisions within them, to keep his scientific interests sated. Another way to think about this kind of circumscribed environment is through the concept of the "ecosphere"—the regional environment that was the entire sphere of existence for most people before the Industrial Revolution. Preindustrial life centered on the ecosphere and the values of bioregionalism that ecological philosophers now cite as a necessary corrective to the "biosphere" mentality of globalization. Individuals in developed nations with a biosphere mentality expect goods to come from anywhere on the planet: they expect materials to be extracted in one location, parts forged in a second, components assembled in a third, and finished products sold and consumed in a fourth, without regard to the distance between places. The ecosphere or bioregional ethic, seen now in retrospect as the prevailing way of life before industrialism, advocates that consumers and goods should originate in the same ecological region (Buell 143).

Over the course of four letters, White comes to realize the strong potential of this serendipitous method of monography. It began with simple regional records designed to enable detailed migratory reports. But White's instincts push his cataloguing science toward innovation: "For many months I carried a list in my pocket of the birds that were to be remarked, and, as I rode or walked about my business, I noted each day the continuance or omission of each bird's song; so that I am as sure of the certainty of

1. The OED records White's use of "monographer" in 1770 as the first in the language. Later in his own narrative, White advises that "a good monography of worms would afford much entertainment and information at the same time, and would open a large and new field in natural history" (197). This monograph was not to appear until Darwin's *Formation of Vegetable Mould Through the Action of Worms* (1881), a work that does not acknowledge White.

my facts as a man can be of any transaction whatsoever" (117). The key concept here is White's notice of omission in migratory patterns. Not merely the presence of an identifiable species but also its absence become formalized as facts in the *Natural History and Antiquities of Selborne*. Here is a crucial turn in methods of ecological knowledge. Science was accustomed to in-depth study of apparent, observable, material entities, but it had no clear interest in the gaps that are equally important to understanding patterns of species distribution over time, especially in disturbed environments. In effect, his monographic focus provides a crucial trial of the stability and continuity over time that natural history had previously assumed to be inherent in ideas like the great chain of being. This discovery of species absence shifts White's original study of phenology into the modern age, and annual migratory cycles are discovered as shifting and unreliable. Modern phenology has become a central method of ornithologists who study the effects of climate change on migratory patterns.

The widespread appeal of White's chronicle rests partially on his caring and concerned voice for all the creatures of Selborne. Revealingly, he shows more affection for oaks, turtles, and worms than for the "hordes of gypsies which infest the south and west of England" (179). White may be accused of class prejudice, but he is also making an implicit statement about the inherent value of nonhuman inhabitants. Human activity too often destroys the peaceable network of other species in Selborne. Where the oak is felled, the intrepid mother bird is struck dead (11); where hunters are unfettered by regulation, the partridges and red deer become rare or extinct, leaving a "gap" in *Fauna Selborniensis* (22); lowly worms, though despised, are essential to soil health. "Earth-worms," White writes, "though in appearance a small and despicable link in the chain of nature, yet, if lost, would make a lamentable chasm. [. . .] Worms probably provide new soil for hills and slopes where the rain washes the earth away; and they affect slopes, probably to avoid being flooded. Gardeners and farmers express their detestation of worms. [. . .] But these men would find that the earth without worms would become cold, hard-bound, and void of fermentation; and consequently sterile" (196).

White's innovative thinking goes beyond the hierarchical great chain of being that was thought to extend from rocks, at the bottom, to plants, animals, humans, and finally God. Here, environmental stress is evident through the deletions in an interdependent biotic network, a web of nature. White's point falls hard on the ignorance of gardeners and farmers who assume the subordination of other species rather than their equality and inherent value. When nature's economy is violated, surprising imbalances occur and have chaotic effects on the web of life. The loss of the red deer may allow for the advance of brambly undergrowth that chokes the forest. Though the balance of biological complexity is his ethic, White needs a catalyst of disturbance and extinction to recognize the value of biodiversity. Degradation is a prerequisite to ecosystem-level conservation.

In time, White's tone shifts from passing elegies for lost species to more concentrated expressions of awe and fear at the unpredictable weather of the 1780s. A primer for the narrative tone at the end of the chronicle comes when a landslide caused by a sudden massive thaw hurls a "huge fragment" of earth hundreds of yards down a steep slope. Houses, woods, and farm fields are "strangely torn and disordered" by the mysterious event. All witnesses agree "that no tremor of the ground, indicating an earthquake, was ever felt" (222–23). In this and other apocalyptic passages, White offers little speculation as to the cause and makes no reference to biblical Armageddon. He seems to enjoy lingering on the perversity of the incident, providing only an objective account that allows sensation to work its own effect in the individual reader. The narrative gains momentum when White considers the effects of these climatic anomalies on established ecological relationships. He explicitly brings meteorology, the study of the unpredictable or "meteoric," into Selborne's history: "Since the weather of a district is undoubtedly part of its natural history, I shall make no further apology for the four following letters, which will contain many particulars concerning some of the great frosts and a few respecting some very hot summers, that have distinguished themselves from the rest during the course of my observations" (253). This letter, the sixty-first of sixty-six, opens an extended exposition on sublime phenomena noted objectively as temperature and barometrical readings, but also on the psychological effects of unprecedented events in natural history. He never returns to his initial phenological perspective, which assumes cyclical, consistent patterns of deistic design open to the naturalist's observation.

White's language comes to rely on exceptional terms quite foreign to a natural theology based on the balanced economy of nature. The words *paradox, severity, loathsome, amazing, tremendous, extraordinary, portentous, superstitious, strange, prodigious, violent, deluging, convulsed,* and *fierce* enliven the final series of letters (253–68). The four letters that detail sudden and unseasonable extremes of warmth and cold prepare the reader for the last two entries, which detail the atmospheric effects of 1783's Laki volcano eruption in Iceland and the severe thunderstorms that accompanied this catastrophe. White uses these extreme observations rhetorically as well as epistemologically. The ethos established by his early talent for close and patient description is a counterbalance for this new narrative of wild weather. It provides a sense of authorial reliability lacking in the work of more histrionic writers. White feels confident as a respectable member of the scientific establishment, as well as an independent-minded scholar who knows the subject of his monograph better than anyone else.

As a microcosm, Selborne's dynamics make intelligible the movements of a larger natural world. White is eager to learn the lessons of the model, however surprised he may be by its recalcitrance. Sudden, unseasonable changes in temperature determine the biological character of entire years; they are not merely passing inconveniences for human beings. The exceptional winter seasons of 1768 and 1776 impress him with "wild and grotesque" scenes of extreme cold and heavy snowfall (258). These "accidental severities," which occur "once perhaps in ten years," provide knowledge of which plants can withstand extreme cold and which succumb when temperatures go off-kilter (256). Conversely, the summer extremes are notable for their effects on animal populations: "The summers of 1781 and 1783 were unusually hot and dry. [. . .] The great pests of the garden are wasps, which destroy all the finer fruits just as they are coming into perfection. In 1781 we had none; in 1783 there were myriads" (263). This passage provides some of the earliest speculation on the future science of population ecology. White's notes reveal that the demographics of some species, like wasps, are subject to wild vacillation each year, without any obvious reversion to a long-term norm and certainly no annual constancy.

Although White did not have the quantitative tools to unravel the mysteries of population fluctuation, his work effectively acknowledges a problem that the science of ecology would model more than two hundred years later. When today's population ecology takes account of a variable environment over many years, the system often shows nonlinear and emergent properties consistent with chaotic dynamics. One important variable that mathematical modelers are currently attempting to capture is the effect of ecological variation on population and community dynamics (Chesson 253). White has no desire to elide or simplify these chaotic patterns that become apparent when closely observed and recorded in the long term. Though he can provide no answers for why two equally hot, dry summers would result in such different wasp infestations, his posing the question started public inquiry.

The wasp population explosion that summer was upstaged by a literal explosion. On 8 June, the Laki volcano in Iceland erupted. White's description of the impact in Selborne speaks for itself:

> The summer of the year 1783 was an amazing and portentous one, and full of horrible phenomena; for, besides the alarming meteors and tremendous thunder-storms that affrighted and distressed the different counties of this kingdom, the peculiar haze, or smokey fog, that prevailed for many weeks in this island, and in every part of Europe, and even beyond its limits, was a most extraordinary appearance, unlike anything known within the memory of man. By my journal I find that I had noticed this strange occurrence from June 23 to July 20 inclusive, during which period the wind varied to every quarter without making any alteration in the air. The sun, at noon, looked as blank as a clouded moon, and shed a rust-coloured ferruginous light on the ground, and floors of rooms; but was particularly lurid and blood-coloured at rising and setting. All the time the heat was so intense that butchers' meat could hardly be eaten on the day after it was killed; and the flies swarmed so in the lanes and hedges that they rendered the horses half frantic, and riding irksome. The country people began to look with superstitious awe, at the red, louring aspect of the sun; and indeed there was reason for the most enlightened person to be apprehensive; for, all the while, Calabria and part of the isle of Sicily, were torn and convulsed with earthquakes; and about that juncture a volcano sprung out of the sea on the coast of Norway. (265)

Up to this point White has maintained his educated distance from the superstitions of country people and has advanced solid objective reports. He appeals here to Romantic discourse that appreciates the wonder of natural forces and delights in their irreducible mystery. Inspired by the same event, William Cowper noted in his journal, "We never see the sun but shorn of his beams, the trees are scarce discernible at a mile's distance, he sets with the face of a red hot salamander and rises with the same complexion" (quoted in Grattan and Brayshay 128). Alternatively, 1783 was known as the "sand summer" in England because of the lingering atmospheric ash, as described in White's account of the blanched midday sun. The Laki eruption

infused a massive volume of sulfurous ash into the stratosphere, which had
the effect of reflecting some sunlight back into space and cooling the atmosphere; but the ash also dispersed admitted light and created a milky-white
luminosity like a frosted incandescent light bulb. As Laki continued to
pump ash and gases into the atmosphere for eight months, widespread famine, stifling air pollution across Europe, and a particularly severe winter
into 1784 condemned most of Iceland's livestock to death, and one-quarter
of its human population followed. Local parish records across England
from 1783–84 suggest that the accumulated effects of the Laki eruption
killed twenty-three thousand British men and women, which makes it one
of the largest natural disasters to beset modern England. An estimated 120
million tons of sulfur dioxide were emitted, or three times the total industrial pollution of Europe in 2006 (Walker).

The Laki eruption had the kind of apocalyptic qualities that led the religious to believe that they were experiencing a form of divine retribution,
and made secular-minded Enlightenment thinkers question the perfectibility of human society through reason and science. Religious, folk, and scientific perspectives preserved in eighteenth-century periodicals demonstrate
the high levels of anxiety and troubling portents swirling in various social
circles (Grattan and Brayshay 129–32). The virulent heat in July 1783,
the violent cold throughout the winter of 1783–84, and the confusion of the
alternately blanched and ensanguined sun raised serious doubts about the
human ability to understand or control chaotic elements of nature. Rather
than dismiss the "superstitious awe" of country people, White feels on an
epistemological par with them. Even the educated had little scientific explanation for the horribly surreal scenes during those years.

Benjamin Franklin was more analytical. Not knowing whether a volcano
was involved, he called the phenomenon a "universal fog" and forthrightly
rendered the mystery a useful predictive mechanism. If dry summer fogs
were to become a new reality, "men might from such fogs conjecture the
probability of succeeding hard winter, and of the damage to be expected by
the breaking up of frozen rivers in the spring; and take such measures as are
possible and practicable, to secure themselves and effects from the mis-
chiefs that attended the last" (377). Franklin wished to secure a useful indicator from a confusing event, and the lesser ecological effects in America
may have permitted his more stoical reaction.

White allows the chaos of this sublime year to remain mischievous. He
turns to literature to make a lasting image of 1783: "Milton's noble simile of

the sun, in his first book of *Paradise Lost*, frequently occurred to my mind; [. . .] it alludes to a superstitious kind of dread, with which the minds of men are always impressed by such strange and unusual phaenomena" (265). The passage he quotes abuts a description of Satan as the "Arch-Angel ruin'd [. . .] th'excess of Glory obscur'd" (*Paradise Lost* 1.593–94). Having fallen, Satan's full angelic sun is occluded by his moral corruption, and his legions are filled "with fear of change" (1.598). Satan's band of fallen angels organizes in ranks, and they emit "A shout that tore Hell's Concave, and beyond / Frighted the Reign of Chaos and old Night" (2.542–43). The revolution itself is a principle of disorder set against divine cosmic harmony. White's allusion to Milton is suggestive: it figures the ensanguined sun following Laki's eruption as a principle of corruption and error. The Laki eruption cannot be ignored, nor can it be explained away; it is one of the chaotic raw elements of the cosmos. As Milton's Satan has only begun in Book I to cause trouble in the balanced hierarchy of God's creation, White perceives a nagging sense of imbalance and future calamity surrounding these "horrible phaenomena" (265).

Recent volcanic events have stimulated new interest in Laki's 1783 performance. When Eyjafjallajökull (AYE-yah fyat-lah yir-kutl) erupted in Iceland in April and May 2010, its ash filled the atmosphere over Europe in a morphing cloud that covered the United Kingdom, France, Scandinavia, and eastern Europe, and extended as far south as Spain and the heel of Italy. As measured by both the Volcanic Explosivity Index and total ash weight, this eruption was a minor event compared to Laki, but it caused the highest level of air traffic disruption since World War II (*Sydney Morning Herald*). The estimated $200 million per day in lost airline revenue and the major economic disruptions for Europe and its trading partners (primarily in Africa, Asia, and Australia) are stark reminders of a global society's economic reliance on long-distance air transportation (Wearden). An eruption today on the much greater scale of Laki might have fewer repercussions for human health because of better technologies, but its economic effects would be dismembering. Eyjafjallajökull was a reasonably polite reminder of how environmental events disrupt modern business as usual. The chaotic narrative sporadically quashes the sunny ideal of steadily growing economies on a supportive ecological stage.

White's *Natural History and Antiquities of Selborne* is more ecologically complex than critics have allowed. The text indeed has extended passages that epitomize economic balance. But the work is innovative for other reasons. Its use of monography allows a microcosmic vision that is echoed in the eco-system concept that would emerge in the twentieth century. The micro-cosm is subject to damaging changes in composition. White observes extinction due to human activity, which he imagines as gaps in an intercon-nected web of life rather than a hierarchical chain of being. Some of his careful observations raise questions that ecological scientists are still actively researching in the twenty-first century, such as the ecology of cha-otic population fluctuation, the minimum viable populations of stressed species, the impacts of deforestation, and the multivariate dynamics of natural disasters, including thunderstorms, landslides, and volcanoes. White's a priori expectation to observe economy in nature by no means blinds him to the importance of extreme, unpredictable weather and its downstream effects over many seasons and across species. There is no indi-cation in the text that White is particularly disconsolate as a result of his uncertainties, but there is a sense that the phenomena are beyond the state of his science. His epistolary narrative is precocious and should be appraised as an important early work in the comparison of ecologies of bal-ance and chaos.

The Last Man

Mary Shelley's novel *The Last Man* (1826) may appear to be a fiction far removed from White's natural history chronicle. Divided by genre and composed within different cultural climates, the two works nevertheless find common ground in their concern for nature's patterns of disturbance. Both Shelley and White look to the Romantic sublime in nature's chaotic plots. Last men enjoyed a literary vogue after the Indonesian volcano Tam-bora's eruption in 1815 and the economic depression of the 1820s. In 1823 Thomas Campbell published a poem with the same name as Shelley's novel, which Campbell claimed had inspired Byron's "Darkness," a poem written in July 1816 under the dark skies of "the year without a summer." These two lyrics convey the visceral feeling of apocalypse that fell over Europe in those surreal summer months. Byron's work envisions volcanic chaos as a barren landscape:

> The world was void,
> The populous and the powerful was a lump,

> Seasonless, herbless, treeless, manless, lifeless—
> A lump of death—a chaos of hard clay.
> (69–72)

Campbell's "The Last Man" centers on a cosmic disease transferred to humans:

> The Sun's eye had a sickly glare,
> The Earth with age was wan,
> The skeletons of nations were
> Around that lonely man!
> Some had expired in fight,—the brands
> Still rusted in their bony hands;
> In plague and famine some!
> Earth's cities had no sound nor tread;
> And ships were drifting with the dead
> To shores where all was dumb!
> (11–20)

Indeed, Tambora's eruption in 1815 was a worthy inspiration for these apocalyptic visions. It was the largest eruption in recorded history, as measured by the volume of magma expelled—140 billion tons (Oppenheimer 230). The sudden ejection of such a mass of pollution into the higher levels of the atmosphere had gradual global climate effects: it took slightly more than a year for the high-flown ash and gases to form an aerosol veil that darkened European and North American skies the following summer. As with Laki, this global event was more than a chilly inconvenience. Tambora caused crop failures, widespread animal deaths, and subsequent famine. Life on earth in 1816 also had the misfortune of particularly low solar activity, which intensified the volcanic winter.

In the poetry of Byron and Campbell, the effect was figured as ecocidal, driving earth to death in either the deep past or the entropic future. To Byron, earth seemed to have reverted to a primordial form, a "lump" and "chaos of hard clay" unsupportive of life and subject to wicked extremes in the elements—a young, undifferentiated planet. Campbell's vision is centered more on the fate of humans and the postapocalyptic vacuum of life and culture extinguished. The sun's "sickly glare" and the aged earth's "wan" face are images of senescence that speak of entropy as the ruling cosmic force dismantling order, and Campbell's sun and earth resemble the bloody sunsets and milky daylight of a volcanic atmosphere. In both poems, despite the wicked weather of 1816, the effects of ecological chaos are too weird to be contemporary; they must be primordial or futuristic.

Details of these climate extremes and their literary offspring in Byron and Mary Shelley are part of the lore of Romantic scholarship. A unique summer like 1816's must make an indelible impression on the mind, especially if it serves as the inspiration for an author's magnum opus, as it did for Shelley in *Frankenstein*. In the spirit of a theme with variations, Shelley's flint stone for catastrophe in *The Last Man* is again chaotic weather, but in this novel the climate is not the clammy summerless depths of Ingolstadt laboratories or the remote reaches of the Hebrides or the Arctic (or Frankenstein's brain); rather, she develops catastrophe out of humid tropical warmth that is an excellent vector for disease. The novel was written in the confines of Shelley's London apartment after the death of three of her children and Percy Shelley's drowning, and the roman à clef explores the widow's new realities in its characters, events, and climates. Her letters in those years reveal how she felt like The Last Woman, marooned apart from her lost generation. She gathers her circle by reanimating the dead in the forms of Percy Shelley and Byron (in the characters of Adrian and Raymond, respectively), and by transforming the chilly London dampness of February 1824, when she began writing, into a lush, tropical England in the last decades of the twenty-first century. Like the later nineteenth-century Thames Valley catastrophes, which include *After London* and *The Time Machine*, Shelley's vision of her perishing civilization invokes the powers of a wild, witchlike Mother Nature. Shelley implicitly challenges the assumption that global trade and colonialism were healthy endeavors, not only for the British body but also for English ecosystems. Influenced by her knowledge of Thomas Malthus's *Principles of Population*, *The Last Man* depicts the environmental checks on population that undercut philosophies of Enlightenment utopia such as those advocated by her own father William Godwin.

The novel progresses from the classic autobiographical beginning of the hero, Lionel Verney, with the first line, "I am the native of a sea-surrounded nook," to the promised singular resolution: "behold the tiny bark, freighted with Verney—the LAST MAN" (9, 470). Any apparent coherence or order in this seeming A-to-Z narrative is misleading. The frame of *The Last Man*, contained in the preface, introduces a second author of the narrative, an unnamed vacationer who in 1818 discovers the scattered "Sibylline leaves" that he assembles into Verney's linear story. *The Last Man* is a narrative of fused fragments confused by time: the human extinction of the late twenty-first century is assembled from fragments in 1818. This discoverer, a cave spelunker, describes his formative editorial role: "I present the public with my latest discoveries in the slight Sibylline pages. Scattered and unconnected as they were, I have been obliged to add links, and model the work into a consistent form. [. . .] Sometimes I have thought that, obscure and chaotic as they are, they owe their present form to me, their decipherer. [. . .] My only excuse for thus transforming them, is that they were unintelligible in their pristine condition" (6–7). These "obscure and chaotic" fragments of a narrative are assembled in a certain order, one that doggedly pursues coherence and causality, when they essentially have none. In their discovered form they are admittedly "unintelligible," and this outer-frame narrator claims responsibility for causal sense in the unfolding of events, including his temerity in composing "links" between fractured episodes. The novel seems unable to fulfill its own prophecy of human extinction. *Frankenstein*'s doubly framed narrative makes an apparent study of each teller's manipulations and reliability. Shelley's framing in *The Last Man* is less coherent, and therefore more mysterious. One could claim that the preface's sole purpose is to seal off logical objections that the narrative of a last man would have no readers, but her placement of the preface anterior to the agonies of the twenty-first century gestures to a more essential, if enigmatic, role for these initial five pages out of nearly five hundred total. The time inversion might suggest that Verney's story is a prophecy of future England, not a lived event, or that Shelley wishes to fragment linear time in order to question assumptions of the inevitable advance of society. Perhaps when nature itself behaves chaotically, narrative follows.

Narrative chaos has been embraced by literary deconstruction, which looks at narrative dynamics through a lens of nonlinearity and contingency (see Hayles; Parker; Conte; Palumbo; and Livingston). Most of this material comes from twentieth-century literature, and James Joyce in particular is a strange attractor for chaos. Carolyn Merchant has suggested in *Reinventing Eden* that chaos, from a narrative theory perspective, "might posit characteristics other than those identified with modernism, such as a multiplicity of real actors; acausal, nonsequential events; nonessentialized symbols and meanings; many authorial voices, rather than one; dialectical action and process, rather than the imposed *logos* of form; situated and contextualized, rather than universal, knowledge. It would be a story (or multiplicity of stories) that perhaps can only be acted and lived, not written at all" (157–58). Harmony, causation, and coherence are constructed from the disordered elements that make up the original story. Any appearance of order in *The Last Man* is based on an illusory cognitive drive to organize chaos.

This cryptic beginning leaves authorship indeterminate and creates a narrative experiment. It is a literary echo of Charles Darwin's notion that the fossil record was an imperfect chronicle of a perfect story of evolutionary gradualism. Darwin filled in the gaps with narrative speculations on the intermediary forms not recorded in fossils. The "history of the world imperfectly kept, and written in a changing dialect," required some spiffing up for Darwin's gradualism to be true (*Origin of Species* 229). If the existing fossil record reflects the pattern of natural history, the evolutionary narrative is chaotically fragmented, like Shelley's sibylline leaves. These texts are original artifacts, and the patched quilt of a coherent narrative pieced together from their fragments may show the author's attempt to conform to the expectations of the nineteenth-century reading public. Shelley's novel has received much more positive attention from postmodernist scholars than it did from her contemporaries. *The Last Man* received widespread critical appraisal only after a new edition was printed in America in 1965 (Parrinder 66). Reflecting the generally poor critical reception in 1826, one reviewer called the novel "the offspring of a diseased imagination, and of a most polluted taste" ("Review of *The Last Man*"). With today's popular tastes, the ecological valences of disease and pollution may be read to great advantage in this prophetic novel, and its narrative chaos is familiar to modern readers.

It is a complicated tale, with a wandering plot and surprisingly conventional characters, not improved by the sentimental and logorrheic dialogue. The novel's value lies in Shelley's perceptive treatment of a chaotic female-gendered nature, her appreciation of radical contingency in natural history, and the remarkable visions of global warming through globalization, which together exacerbate the spread of disease. Her characters are tortuously, farcically Romantic, but the imagined twenty-first-century climate she describes is eerily apt. Frederick Buell's analysis of apocalypse reminds us that "plague" has dated, almost medieval connotations for individuals in modern developed nations, who often believe that medical technology and inoculation have eradicated epidemiological threats to our bodies (132). However, Shelley realizes that disease transmission will only increase as the climate warms. Shelley's novel shows a perverse reversal of the colonial project by depicting the decline of the British body as it is colonized by exotic microbes, and an assault on English nature by advancing tropical species.

Where *Frankenstein* drew scenes of sublime terror evoked by the vast arctic plains, ending with the blind image of the creature "lost in darkness and distance," *The Last Man* capitalizes on the paradoxical horror of a too-pleasant nature mocking psychological despair. The early arrival of the warm season indicates the arrival of the survivors' annual trial by plague. Mother Nature reveals her vindictive, witchlike properties in the face of humanity's reasoned opposition:

> Nature, our mother, and our friend, had turned on us a brow of menace. She shewed us plainly, that, though she permitted us to assign her laws and subdue her apparent powers, yet, if she put forth but a finger, we must quake. She could take our globe, fringed with mountains, girded by the atmosphere, containing the condition of our being, and all that man's mind could invent or his force achieve; she could take the ball in her hand, and cast it into space, where life would be drunk up, and man and all his efforts for ever annihilated. (232)

This willful, vindictive, powerfully destructive characterization of Nature, the portrait drawn in chaos ecology, was originally embodied as the fallen Eve in the Western tradition (Merchant, *Reinventing Eden* 157). This gendering of Nature reawakens mythological traditions of natural power lying in the laps of personified goddesses, and Shelley extends the accusatory "she" from natural climate to the pestilence itself as a she-disease. The war between the sexes is fought along conventional lines, the revolting element of feminine nature, climate, and disease pitted against masculine human culture, science, and reason. As in *Frankenstein*, there are no female human characters with intellect or agency, though females have a disproportionate share of emotive dialogue. Nature, however, is a feminine element with chaotic agency that overcomes masculine reason, yielding a chic element of proto-ecofeminism to Shelley's work (McKusick 109).

The tone in the passage quoted above echoes a Malthusian worldview, which helped Charles Darwin envision how survival itself was a virtue that affected evolutionary progress. Malthus's "Essay on the Principles of Population" (1798) proposes that the plight of human experience (war, famine, disease) could not be wholly extirpated by Enlightenment institutions such as democratic government, intensive technological farming, and enhanced medical technology. Shelley's father, William Godwin, an Enlightenment political idealist, wrote a lengthy refutation of Malthus's essay. However, Mary Shelley's novel consistently builds and then systematically destroys schemes of Enlightenment-rational and Romantic-imaginative hope developed by her male characters (Paley xv). *The Last Man* is a Malthusian work without recourse to salubrious progressive evolution. In describing the capricious moods of nature, Shelley figures Mother Earth as the author of both jeremiads and idylls; her duplicity is all the more unsettling. For chilled British readers, there is a notable irony that a warmer world resembling colonies in the British Indies (both East and West) may dress up like paradise, but the climate change is an epidemiological nightmare. Shelley takes the Malthusian notion one notable step further by envisioning a world in which even Edenic, productive, and nurturing Nature offers no succor to the cursed human race. Much worse than providing a challenge to survival, Lionel Verney comes to know the pleasant natural world as a set of false signs that belie a fate of death by disease. Order and balance seen in nature are illusive hopes, manifestations of overwrought human cognition rather than a true mirror of larger intelligible forces at work in the cosmos. Verney's narrative repeatedly returns to microcosmic images of order and containment lost to the catastrophe of universal human decline.

Part of this lost control over the ecological world seems attributable to the lost cultural control of the imperial power. Alan Bewell's *Romanticism and Colonial Disease* devotes a long and thorough chapter to *The Last Man*'s epidemiology. England harbors ships that have landed on every continent, bringing their stores of trade goods, humans and other animals, and microbes. Each one of these categories has a quality of invasiveness to it. Goods from other climates invade the British identity and change consumer appetites, a trend particularly suspect in the nineteenth century, with the import of addictive opium and foods reliant on slave labor (coffee, sugar, chocolate). Foreign men contribute to the worldliness of London, but their settling in England provides an early glimpse of immigration debates, which often took a pointedly racialized tone. Globalization allows the easy traversal of disease from more resistant populations to more vulnerable ones. Colonial history shows that most diseases were carried from Europe to the native populations of colonized lands, but in Shelley's novel this pathway is reversed. The plague arrives in England from an American trade vessel, and in Italy from the nearest Eastern country: Turkey. Europe is laid open to the world's diseases as a porous, susceptible body.

Part of the novel's paranoia must be accounted for by the very real concerns about urban sanitation in the nineteenth century. Cholera arrived in England six years after the publication of Shelley's novel, during the second cholera pandemic. The first pandemic affected India, China, and Indonesia, including British colonial regions where there was a large military presence (Paley xiii). The second pandemic reached London and Paris in 1832 from its origin in the Ganges River Delta, and it claimed sixty-five hundred lives in London and a hundred thousand in France (Rosenberg 101). Cholera remained a serious water-borne threat in Europe until 1851, the year of Shelley's death. Her generation directly experienced how global trade routes served as disease vectors that could rapidly and efficiently carry bacteria from locales of origin into vastly different climates and populations, moves that often dramatically increased a disease's virulence.

Shelley's depiction of the rapid spread of disease in the mysteriously warmer English climate of 2100 aligns with present-day epidemiological concerns about how pathogen habitats will be expanded via climate change. As Frederick Buell notes, environmental despoilment in the twenty-first century involves a constant network of exchange among macro-, meso-, and microbiological conditions: "Raising the likelihood of a substantial increase in serious infectious disease are trends like climate change, development, habitat destruction, pollution, overpopulation, urban slummification, the industrialization of agriculture, and the rise of global transport and mobility. A host of decisive human modifications of natural and social environments— rapidly expanding modifications that lead not just to the destruction of macroecosystems but also to deeply problematic alterations in microbiological environments, are thus responsible for the rise of infectious disease" (129–30). This twenty-first-century perspective shows how the ecology of disturbance operates at many scalar equivalencies simultaneously—the macrocosmic issue of global epidemiology must consider microcosmic conditions in local climates and the immune capacities of local populations. Even as antibiotics have dramatically curtailed individual suffering from infections, they have the epidemiological side effect of promoting mutant bacteria. Liberated from intraspecies competition, our antibiotics select survivor microbes, or "superbugs," that become the next generation to infect human populations.

Nineteenth-century theories of miasma suggested that unhealthy environments had certain characteristics, particularly fog and dampness that incubated "bad air." Miasmatic theory was used to explain the cholera outbreaks in large European cities and provided the basis for major renovations that cleared out stagnant waterways in urban areas and drained wetlands in the country. These measures did indeed improve the sanitary and health situation, but not for the reasons miasmic theory cited. When John Snow discovered in 1854 that the epicenter of the London cholera epidemic had been the Broad Street pump in Soho, cholera transmission was correctly linked to waterborne germs, and the ground was laid for the identification of microbes as the causes of illness. Shelley's treatment of Lionel's inoculation certainly relies on the contemporary "bad air" conventions of miasma, but the scene is much more complex than the cliché of the damsel catching a chill from her evening walk in the moors. He enters a dark room in London: "A pernicious scent assailed my senses, producing sickening qualms, which made their way to my very heart [. . .]. I lowered my lamp, and saw a negro half clad, writhing under the agony of disease, while he held me with a convulsive grasp. With mixed horror and impatience I strove to disengage myself, and fell on the sufferer; he wound his naked festering arms round me, his face was close to mine, and his breath, death-laden, entered my vitals" (336–37). Here, it is not London's bad breath that is infectious but the racialized encounter that throws the Englishman into the arms of the African, in a gust of colonial breath and tropical disease. Plenty of racial anxiety is revealed through the quasi-intimacy of their entanglement; this could not be an arbitrary choice of disease vector. This scene speaks to British anxiety about the effects of colonialism, which is the most popular interpretation of the novel. What is missing from these colonial and epidemiological readings is an account of how a disturbed nature provides the stage for this drama. There would be no disease exchange in this encounter between racial others if the climate had not already begun to shift away from established patterns.

The unnatural global warming estranges Europeans from their environments of adaptation, heightening their susceptibility and bringing them into an intimate common fate with the world's other peoples. Contagion makes the other into brother. *The Last Man*'s global warming theme begins with descriptions of war-torn Constantinople, gateway between East and West:

> The southern Asiatic wind came laden with intolerable heat, when the streams were dried up in their shallow beds, and the vast basin of the sea appeared to glow under the unmitigated rays of the solsticial sun. Nor did night refresh the earth. Dew was denied; herbage and flowers there were none; the very trees drooped; and summer assumed

the blighted appearance of winter, as it went forth in silence and flame to abridge the means of sustenance to man. [. . .] All was serene, burning, annihilating. [. . .] The sun's rays were refracted from the pavement and buildings—the stoppage of the public fountains—the bad quality of the food, and scarcity even of that, produced a state of suffering, which was aggravated by the scourge of disease. (189–90)

Here, the disease is secondary to the famine caused by the failure of the rains. The climate's shift toward a desertlike ecosystem has occurred too rapidly for flora or agriculture to adapt, and the plague's path cuts directly through the bodies of a weakened population. Before long, both the climate and its sidekick, the plague, have swept across Europe into England. This is the point at which the other disorders of civilization, war and colonialism, fall into inconsequence compared to the "eruptions of nature" (232). The people start to balk at catastrophe: "Can it be true [. . .] that whole countries are laid waste, whole nations annihilated, by these disorders in nature?" (233). Orators mislead their countrymen into believing that the English are not subject to the natural disasters as natives of the tropics are:

> Countrymen, fear not! [. . .] [Plague] is of old a native of the East, sister of the tornado, the earthquake, and the simoom. Child of the sun, and nursling of the tropics, it would expire in these climes. It drinks the dark blood of the inhabitant of the south, but it never feasts on the pale-faced Celt. If perchance some stricken Asiatic come among us, plague dies with him, uncommunicated and innoxious. Let us weep for our brethren, though we can never experience their reverse. Let us lament over and assist the children of the garden of the earth. Late we envied their abodes, their spicy groves, fertile plains, and abundant loveliness. (233)

This speech, with its sneering blend of false pity and racial pride, screams for correction. It comes on the very next page, where those English who might have envied the "spicy groves, fertile plains, and abundant [ecological] loveliness" of the tropics find, to their dismay, their wish granted (237). In August, the disease in an oddly hot England "gained virulence, while starvation did its accustomed work. Thousands died unlamented; for beside the yet warm corpse the mourner was stretched, made mute by death" (235).

Several deadly years on, the four English seasons have fallen completely off
their orbit:

> Winter was coming, and with winter, hope. In August, the plague
> had appeared in the country of England, and during September it
> made its ravages. [. . .] The autumn was warm and rainy: the infirm
> and sickly died off—happier they. [. . .] The crop had failed, the bad
> corn, and want of foreign wines, added vigour to disease. Before
> Christmas half England was under water. The storms of the last win-
> ter were renewed. [. . .] But frost would come at last, and with it a
> renewal of our lease of earth. Frost would blunt the arrows of pesti-
> lence, and enchain the furious elements; and the land would in spring
> throw off her garment of snow, released from her menace of destruc-
> tion. It was not until February that the desired signs of winter
> appeared. For three days the snow fell, ice stopped the current of the
> rivers, and the birds flew out from crackling branches of the frost-
> whitened trees. On the fourth morning all vanished. A south-west
> wind brought up rain—the sun came out, and mocking the usual laws
> of nature, seemed even at this early season to burn with solsticial
> force. It was no consolation, that with the first winds of March the
> lanes were filled with violets, the fruit trees covered with blossoms,
> that the corn sprung up, and the leaves came out, forced by the unsea-
> sonable heat. We feared the balmy air—we feared the cloudless sky,
> the flower-covered earth, and delightful woods, for we looked on the
> fabric of the universe no longer as our dwelling, but our tomb, and the
> fragrant land smelled to the apprehension of fear like a wide church-
> yard. (269–70)

The plague is borne of ecological disturbance first and foremost. The pleasing signs of spring come unearned by the privations of winter, and the
missed cycle of cold allows the plague to overwinter without check. The violets, fruit trees, and corn still have English ecological origins, but their
rhythms are spun into strange oscillations as the warmer months gain dominance, making evolutionary shifts in species composition inevitable.

This estrangement between humans and their accustomed environment
threatens psychological consequences that we are beginning to understand
in this age of climate change. What dies is the cognitive comfort of the
home ecology, the well-known *oikos* that carries yearly rituals and passes

time in its intelligible, cyclical way. The psychology of disturbance inherits this vacated brain space, as it did for Mary Shelley dwelling beneath the parasol of Tambora's ash in the lost summer of 1816. With a "global weirding" mentality, we come to expect weather anomalies and look to anthropogenic sources to explain them.[2] Chaotic tangles of human and natural activities seem to underlie every event, even the pleasantly warm day in winter. We become less able to enjoy ourselves within the weather, whatever its conditions, because they seem to be signifiers of an inexorable process already on the move, a beast slouching toward Bethlehem.

This beast is borne out by the frequency of objectively measured landmark weather events. As of this writing, the twenty warmest years in the last 130 (when the National Climatic Data Center began measurement) have all occurred since 1983, and every year since 1977 has been above the average set during that 130-year period (NCDC). Habitat shifts result from climate shifts: animal species are migrating about four miles every decade toward the poles, and average temperatures are moving much faster, at thirty-five miles pole-ward per decade (Hansen 146). That is, as each decade passes you need to move thirty-five miles north in the Northern Hemisphere, and south in the Southern Hemisphere, to experience the same average temperatures. At this rate, New York City would have present-day equatorial temperatures in about seven hundred years, and London in just over one thousand years, with summer seasons much more severe than at the equator because of the tilt of the earth's axis.

Species are also migrating uphill for cooler conditions, which crowds out the former flora and fauna of the highlands. Mary Shelley's fearsome vision is both ecologically and psychologically prescient. Disease may have been the most convenient and apparent cause of human extinction in her fictional apocalypse, but standing back from the sensations of the plague and colonialism for a moment, we can see that a disturbance in climate is the real baseline point of engagement in this novel.

What mechanism causes the earth's warming in *The Last Man*? The novel is silent on this point. The narrative is much more detailed in its description of events as they occur than in any hypothesis of a cause. For a vision of the late twenty-first century, Shelley's work is surprisingly poor in futuristic

2. Coined by Hunter Lovins, co-founder of the Rocky Mountain Institute, the term "global weirding" is an alternative to "global warming," reflecting the belief that climate change causes various weather-related extremes, including both hot and cold weather, to become more intense and to occur out of season.

detail: there is little advance in technology, industry, or political or social life. With its lexicon stuck in early nineteenth-century conventions, *The Last Man*'s most prescient features are its disturbed landscapes and altered climate patterns. In a dream vision, a mainstay of the Romantic imagination, Lionel is haunted by a scene of a great feast turned foul, where goblets "were surcharged with fetid vapour" and his friend Raymond, "altered by a thousand distortions, expanded into a giant phantom, bearing on its brow the sign of pestilence. The growing shadow rose and rose, filling, and then seeming to endeavor to burst beyond, the adamantine vault that bent over, sustaining and enclosing the world" (202). Meager though its meaning may seem, this image—half disease, half pollution—is reminiscent of a volcanic eruption and the darkness that follows. The convex sky that shelters life from vacuous outer space is infected from within. Greenhouse gases are the fetid vapors that would push the landscape toward the imagined climate of 2100. Twenty-first-century culture, familiar with climate disturbance, has come to appreciate the apocalyptic vision that makes *The Last Man* a classic. Shelley's secular apocalypse became a popular convention in later nineteenth-century fiction, laying a foundation for the two Victorian works discussed in the next chapter, Richard Jefferies's *After London* and H. G. Wells's *The Time Machine*.

Shelley's work is a dirge for her portion of the Romantic period. But even with this funereal, backward-looking exigency, Shelley creates something new in her vision of human fate, and out of a deep personal sadness brings forth a text that arrives at a new way of knowing the world. Her fictional proxy Lionel Verney is touched by an excerpt from *Macbeth*, which he hears at a London theater during the plague years: "Alas, poor country, almost afraid to know itself. It cannot be called our mother but our grave, [. . .] where violent sorrow seems a modern ecstasy" (4.3.164–70). Shelley's sorrow is modern because her vision has fused into a valid ecological forecast in this, her future world.

The modern ecstasy of violent sorrow has become a big business in entertainment, with apocalyptic films and novels using the sensationalism of chaos for its thrill value. It is a kind of ecstasy characteristic of a complacent society to pay for a seat within a cool dark room and witness a marvelous spectacle of the world destroyed by fire, flood, drought, hurricane, or disease, even as these forces become more virulent outside the theater. Mary Shelley found her audience in our time, and *The Last Man* echoes through recent dystopian films such as *Children of Men* (2006) and *Contagion* (2011).

2

The Science of Chimneys and Calderas

The atmosphere started absorbing entirely new levels of greenhouse gases beginning with the Industrial Revolution in late eighteenth-century Europe, and levels have been on a nearly exponential increase ever since. The recognition that industrial air pollution had the power to change landscapes, species composition, and human health came to public notice with increasing concern over the course of the nineteenth century. Ominous images like Blake's "dark Satanic Mills," Dickens's London sky shedding "soft black drizzle, with flakes of soot in it as big as full-grown snow-flakes," Gaskell's northlands factories "puffing out black 'unparliamentary' smoke," Ruskin's "two hundred furnace chimneys" vomiting something that resembles "dead men's souls"—these images evoke the aura of nineteenth-century urban British literature that writers alternatively fetishized and fled from to the green countryside (Blake 153; Dickens 1; Gaskell 55; Ruskin 637). Before taking an in-depth look at two Victorian science fiction novels that use industrial catastrophe as the creative energy that forges a new landscape, it will be useful to see how Victorian advances in atmospheric science informed contemporary thought.

Over the course of the nineteenth century, the scientific investigation of air pollution from volcanoes and factories greatly improved the understanding of local and global interconnection. Atmospheric chemists came to understand the composition of smoke by studying volcanic eruptions, which were known to cause human health problems, strange corrosive rains discovered to be acidic, and climate aberrations for years afterward. Earth could be seen as a biosphere, with an insulating layer of ozone and a force field of gravity that held in the atmosphere of the closed global system. Although this cozy arrangement often inspired praise of divine design or just wonderful cosmic luck, other implications of the closed cosmos became apparent, such as the possibility that air pollution does not simply vanish; it insidiously accumulates above.

The first scientific theory that can be identified with modern global warming came from the experiments of Joseph Fourier, a French physicist who pursued a theory of heat conduction. Chaos philosophers Ilya Prigogine and Isabelle Stegners identify Fourier's work with the first conceptions of complex nonlinear systems, the science of complexity (104–5). His "General Remarks on the Temperature of the Terrestrial Globe and the Planetary Spaces" (1824) envisions the earth as a giant greenhouse. Fourier's scheme of natural atmospheric insulation is cosmically benevolent: the gases and water vapor that collect at the outer reaches of the earthly sphere provide essential incubatory warmth for the plant and animal life on the surface. While the gases emitted by human industrial activity were identical to those naturally occurring in the stratosphere, Fourier did not pursue a theoretical connection. The earth's system simply seemed too large, and the volume of greenhouse gas released by humans, too modest. Fourier also had the reassurance of religious faith, which reinforced a theory of active divine benevolence.

In 1861 John Tyndall furthered Fourier's ideas by demonstrating the high absorbent power of gases in the atmosphere, including carbon dioxide and ozone. His conclusions, like those of his predecessors, tended to place value on the insulating power of naturally occurring ozone gases, which seemed to keep the ice ages of deeper geological history at bay. Though local pollution in industrial centers was a health concern by midcentury, an overall trend of warming seemed blithely utopian to the chilled blood of northern European scientists. The term "greenhouse effect" was coined by the University of Wisconsin professor Thomas Trewartha in 1937. His *Introduction to Weather and Climate*, written more than a century after Fourier theorized the global greenhouse, is among the earliest scientific studies to raise alarms about rapid increases in greenhouse gases. The intervening century had rendered great changes in the atmosphere, but they were not nearly as significant as the changes wrought in the decades since Trewartha's work.

Residents in England's newly industrial cities, particularly Manchester and London, bore witness to their polluted microclimates. Elizabeth Gaskell provides a fictional account of the city of Milton in her novel *North and South* (1855): "For several miles before they reached Milton, there was a deep lead-coloured cloud hanging over the horizon in the direction in which it lay. It was all the darker from contrast with the pale grey-blue of the wintry sky. [. . .] Here and there a great oblong many-windowed factory stood up, like a hen among her chickens, puffing out black 'unparliamentary' smoke, and sufficiently accounting for the cloud which Margaret had taken to foretell rain" (59). The 1844 legislation meant to curb industrial emissions went largely unheeded by libertarian factory owners, among them Gaskell's industrialist John Thornton, a Romantic hero for Victorian times. By 1866, Manchester's medical representative identified the citizens as among the unhealthiest in Britain as a consequence of inhaling the air (Christianson 21).

The Scottish chemist and industrial critic Robert Angus Smith discovered acid rain in 1852 in the environs of Manchester. He published the long-researched monograph *Air and Rain: The Beginnings of a Chemical Climatology* in 1872. Acid rain had the alarming power of dissolving the features of building façades (notably gargoyles), faces already shrouded in carbon precipitate. Since the beginning of the nineteenth century, industrial smoke, and particularly sulfur dioxide, has decreased the pH of rain from a neutral 6 to a marginally acidic 4.5 or 4. Readings of 2.4, the acidity of vinegar, have occasionally been recorded in heavily industrialized areas ("Acid Rain"). Events of *Waldsterben*, the death of trees, had been known to follow large volcanic eruptions, but forest dieback near industrial areas in England and the Black Forest in Germany came to be linked to acid rain. Acid rain's tendency to deface tombstones and public statues is a particularly bracing example of the self-annihilating side effects of industrial emissions. Even names carved in stone were imperiled by the airborne appetites of sulfur dioxide and oxidized nitrogen compounds.

By century's end, scientific consensus showed the effects of Manchester's pollution on its plant life, humans, and habitations. In an 1893 essay called "The Air of Large Towns," Manchester chemist G. H. Bailey made an emotional appeal to industrialists about their urban air filters, the trees. In his discursive preamble to the scientific data on pollutants, he wrote, "General experience has shown that evergreens cannot be grown in the heart of our larger cities and even the more hardy deciduous trees make little progress and sooner or later succumb. The sulfurous and other noxious vapors and the deposits of soot, hydrocarbons, etc., which form on the leaves are the chief agents in the destruction of plant-life" (201–2). Trees are only a step removed from animals, and Bailey invoked Britain's famed urban fogs, "when the air is supercharged with such impurities," to convey the pathos of the "fog demon," a predator on public health in the new industrial environment: "The death-rate indeed from such [respiratory] diseases after foggy weather frequently increases to three-fold its normal value and is always exceptionally high in the densely populated districts" (202). Although urban centers in England had long been the loci of infectious disease transmission, Bailey's essay identified a new health risk to industrial city dwellers that was not restricted to humans. Second natures, the patchy clusters of trees that serve aesthetic needs in cities, have their own version of black lung. Their sickness is a poignant symbol of the transorganismal health effects of air pollution. In 1910 T. S. Eliot lyrically described how the urban yellow fog "rubs its back upon the window-panes [. . .] / lingered upon the pools that stand in drains [. . .] / curled once about the house, and fell asleep" ("Prufrock" lines 15, 18, 22). The "fog demon" had become endemic, another offspring of industry.

Victorian factory architecture provides another measure of industrial air pollution. The statistics on average chimney height over the course of the nineteenth century reveal how legislation attempted to curtail urban pollution without the bother of regulating emissions. Gale Christianson traces the growth of "Cleopatra's needles" from an average somewhat below 300 feet in the early decades of the century to new records of 435.5 feet in 1841 and 454 feet in 1857. An average of one hundred chimneys rose each year in London between 1846 and 1853. The House of Commons Select Committee on the Smoke Nuisance, created in 1843, recommended that manufacturers be removed from the city center to a radius of five or six miles (Christianson 56–59). These measures substituted a visibly apparent local environmental problem for an almost invisible but widespread trend toward the blackening of England, urban and rural. As these industrial forests came to define city skylines, architecture trended toward more attractive and ever-taller stacks capable of distributing their effluent over a wider area. Aesthetic chimneys included those designed after the Egyptian prototype, Cleopatra's needle. Perhaps it helped with public opinion to have their aesthetics contribute some artistic character to otherwise utilitarian cities.

Industrial melanism is a case study in evolutionary ecology, and it illustrates this transformative period in England's history. A mutation for uniform darkness in the peppered moth, *Biston carbonaria*, helped the rare species to achieve newfound competitive success with its rival, *B. betularia*. The *B. carbonaria* mutant was better adapted to industrial nature: its color matched the sooty trunks of trees, and predatory birds targeted the more apparent *B. betularia*, whose mottled wings had evolved to blend in with trunks spotted by white lichens. Lichens are good indicators of air quality, and they had dwindled in the English countryside over the first century of industrialism. By 1895, the overwhelming majority of peppered moths in English forests were the *B. carbonaria* variety, a reversal of the original preindustrial distribution. This case is among the first cited examples of natural selection in action, and the mechanism is especially interesting as evolution driven by industry. When legislation provided more effective regulation of industrial emissions and air quality improved toward the end of the twentieth century, the lichen look-alike mottled *B. betularia* came to predominate once more in English forests.

Atmospheric sciences originated from a combination of Enlightenment chemistry, which made great advances is elucidating the chemical nature of air, and the geological explorations of volcanoes at the turn of the nineteenth century. Humphry Davy, Alexander von Humboldt, and James Smithson scrambled around the calderas of the world's most active volcanoes in pursuit of applied information and material samples. Davy climbed Mount Vesuvius fourteen times in 1820 alone, inspired by his research in coalmines to test the various contemporary theories on volcanic action (Matthews 197). Measures of volcanic emissions, both quantitative and qualitative, became crucial data used to parse many competing theories on the role of volcanoes and earthquakes in the earth's history. Alexander von Humboldt's adventurous research on South American geology, which formed the basis for early theories of species distribution and biogeography, also relied on observations of live eruptions. Humboldt developed the "cyanometer" in order to quantify the blueness of the sky. He measured the notably dun skies around active volcanoes, which had the additionally sublime effect of coloring the surrounding objects in lurid, hypernatural hues. He recognized volcanic action as the key to opening new theoretical schemes in geophysics: "Volcanic phenomena [. . .] considered in the totality of their relations, are among the most important topics in earth Physics. Burning volcanoes appear to be the effect of a permanent communication between the molten interior of the earth and the atmosphere that envelopes the hardened, oxidized crust of our planet. [. . .] [Volcanoes provide information on] that intimate connection between so many diverse phenomena" (quoted in Sachs 42). A deep time ruled by catastrophe captured the literary imagination as well: volcanoes, earthquakes, and comets emerged as chaotic, world-ruling events, perhaps as substitutes for more conventional deism. In 1822 Byron raised catastrophic speculation to the level of a new, atavistic mythology: "Who knows whether, when a comet shall approach this globe to destroy it, as it often has been and will be destroyed, men will not tear rocks from their foundations by means of steam, and hurl mountains, as the giants are said to have done, against the flaming mass?—and then we shall have traditions of Titans again, and of wars with heaven" (quoted in Palmer 56). Byron's wish for this sublime, technogeological battle reveals the imaginative energy contained in catastrophe science, which also glosses Byron's revolutionary desire for the old realm to be forged anew.

The comprehension of volcanic eruptions led atmospheric chemists to draw an analogy with industrial pollution. Like eruptions, industrial pollution caused climate anomalies linked to the release of aerosols and greenhouse gases. (Inversely, volcanic eruptions generally cause rapid cooling because of the predominance of parasol-like aerosols that they pour into the stratosphere, and industrial pollution forces gradual warming because of the sheer volume of greenhouse gases that gather in the lower troposphere.) These theoretical connections, though not fully explicit until continental drift theory unified geophysics under a new paradigm in the twentieth century, were common to scientific discourse from the beginning of the nineteenth century. In 1804, the editors of the *Edinburgh Review* marveled at the inherent soundness of an initial link between volcanic and industrial pollution: "it is wonderful how it so long eluded observation, when the slag of every furnace exhibits [the relationship] in the most striking manner" (quoted in Matthews 197).

The analogy between industrial pollution and volcanoes inspired reasonably sophisticated work from scientists. The volcanic ash left behind in the lower geological layers was evidence of ancient volcanic activity, which might support or refute the plutonist theory of geology. Plutonists claimed that volcanic activity was the primary creative geological force in earth's history, and the overwhelming proportion of igneous rocks (cooled from lava and magma) in the earth's crust was good evidence that volcanoes were essential to understanding geomorphology. Chemical analyses of volcanic emissions, both ancient and modern, provided quantitative details about the molecular cocktail in the ash. James Smithson, whose name is preserved in Washington's Smithsonian Institution, produced his analysis of Vesuvius soot in a paper for *Philosophical Transactions* in 1813 in the hope of supporting plutonism. The chemical complexity of these fiery by-products was only beginning to find the light of empirical science, and with a blowpipe analysis Smithson found at least "nine species of matters" in his analysis: "Every thing tells that a large body of combustible matter still remains closed within this stony envelope, and of which volcanic eruptions are partial and small ascensions. Under this point of view, an high interest attaches itself to volcanoes, and their ejections. They cease to be local phenomena; they become principal elements in the history of our globe; they connect its present with its former condition; and we have good grounds for supposing, that in their flames are to be read its future destinies" (quoted in Ewing 74). Earth's future destinies can be read especially well in volcanoes when the same chemicals are released from smokestacks. The identifiable compounds Smithson precipitated out of these Vesuvian salts are as follows, in decreasing abundance: sulfates of potash and soda, muriates of soda, ammonia, copper, and iron, and miscellaneous metallic submuriates. The arcane chemical terminology used here obscures the direct connection with later analyses of industrial pollution. Still, the chemical link remains: sulfates are the salts of sulfuric acid, and muriates are derivatives of hydrochloric acid such as potassium chloride. Eighty years later, in an analysis of air particulates in smoggy Chelsea, G. H. Bailey found the chemical composition at 4.3 percent sulfuric acid, 1.4 percent hydrochloric acid, 1.4 percent ammonia, 2.6 percent metallic iron, and 31.2 percent "other mineral matter," particularly silica (sand) and ferric oxide (oxidized iron) (201). Carbon and hydrocarbons made up the remaining 60 percent. In chemical terms, the particulates emitted by volcanoes over geological time share their molecular structure with the new pollutants of industry, particularly elemental carbon, sulfuric acid, hydrochloric acid, and ammonium. Chemistry changed dialects over the course of the nineteenth century, but it nonetheless became clear that sensational events in geological history informed an anthropogenic future.

The relationship between global climate and any source of atmospheric destabilization is complex. Because of the reflective effect of high-flung volcanic dust particles, especially sulfur compounds that reach the stratosphere, volcanic pollution usually has a parasol effect that causes global cooling for as long as the particles are suspended, sometimes for several years. Smaller eruptions that release a greater proportion of sulfurous gases but with less overall ash can cool the climate dramatically more than larger eruptions, so the chemistry of the eruption is essential to predicting the extent of global cooling. Tambora's eruption in 1815 caused 1816's cold summer in Europe and North America, and Pinatubo's eruption in 1991 depressed global temperatures by 0.6 degrees Celsius during 1992 and 1993 (Harpp 2). It is possible that volcanic activity as illustrated in the Pinatubo effect masks the extent of global warming from greenhouse gases. The cocktail of sulfur dioxide, hydrogen chloride, and hydrogen fluoride released during eruptions combines with water in the atmosphere and falls as acid rain. Three-quarters of the livestock in Iceland succumbed to the combination of frigid temperatures and acid rain exposure after Laki's eruption in 1783 (Harpp 1). Volcanoes also emit carbon dioxide and other greenhouse gases, like anthropogenic pollution, which can have a warming effect. Their emissions today are trivial compared to human production of greenhouse gases: on an annual basis, volcanoes emit one ten-thousandth of the carbon dioxide that human activity does (Harpp 1). Still, the scientific investigation of volcanic impacts on climate is a serious pursuit for atmospheric scientists seeking to understand how chaotic climate patterns develop in response to aerosols and greenhouse gases.

During this era, ecological damage was becoming increasingly apparent, as the expansiveness of industrialism adversely affected air quality. As the tangling of ecological destruction and industrial innovation beckoned new literary beginnings, the modern landscape became a focal character around which to organize futurist fiction. Literary works of ecological chaos are auguries of the state of modern environments as polluted, fragmented, and unstable. As we proceed from Romantic to Victorian visions of nature, the locus of responsibility for disturbance shifts from nature's sublime wildness to industry's bewilderment of nature. This bold shift from nature as a stage for human actors to nature as an actor in the evolutionary drama both naturalizes humanity and humanizes nature. The great insight of nineteenth-century biology, that humans share a common evolutionary origin with all earthlings, is the first step taken in a modern literature of nature. The second step is the admission that humans share a common fate with those other earthlings.

After London, or Wild England

Richard Jefferies's novel *After London* (1885) imagines a postapocalyptic Great Britain, not immediately following the calamity but about four generations "after London ended" (11). Jefferies gives nature a full century of fallow time wherein the landscape evolves beyond the moment of disturbance. In a general way, Jefferies's vision can be read as a continuation of Shelley's drama of human decline. Both novels are proleptic, both center on the Thames Valley, and both use plague as a check on human populations. The most important difference between them is that Jefferies is sensitive to the pernicious effects of industrial pollution on London and its environs. With a talent for describing ecological succession in fallow lands, Jefferies develops an innovative narrative about the chance-driven process of species surviving catastrophe and radiating to fill newly opened niches—what ecologists call the founder effect. Although ecological succession toward a climax community would become a gradualist ideal asserted by Frederic Clements in the early twentieth century, the catastrophe itself is a sudden ecological punctuation with downstream effects in evolutionary time.

Jefferies's vision is a forceful reversal of the trends toward landscape development in nineteenth-century England. From a highly altered second nature riddled with industry, sprawling cities, and new agricultural claims, the novel narrates a successional recovery of first Nature, a move backward toward neowilderness. Surviving humans live in an intrinsic intimacy with their environment that had been lost to their industrial-era ancestors. Emotional intimacy with this primary Nature is based both on its grandeur and on the learned ability to survive as a pioneer in the novel's "back-to-the-earth" ethos. Where modern environmental philosophers advocate extending an ethic of care and emotional attachment to the degraded environments in which we live, Jefferies imagines this commitment as possible only by a return to primary Nature. In this way, his work correlates with the American Romantics and their more recent descendants, who advocate an ethic of wilderness preservation, and with the ectopias of other Victorian writers like William Morris and Samuel Butler.

However, Jefferies's work is innovative because primary Nature coexists with the pernicious ghosts of industry past. His novel's chivalric knight is valorized by surviving the fire dragon of the toxic-waste sites in old London. This deep ecological vision of primary wilderness set against the corruption of dead industry evokes a crisis of modernity common in Jefferies's writings (Hooker 38–43). Revolutionary natural forces provide a counterbalance to the industrial disturbances of modern society. Most engagingly, the liberated character of Nature becomes the prima donna on the primordial postindustrial stage. *After London* is sparsely peopled, and its characters, including the hero, Felix, occupy inherently secondary and passive roles compared to the great natural forces surrounding them.

Living in the age of Darwin, Jefferies benefited from the insights of evolution by natural selection without fully subscribing to a Darwinian worldview. In a notebook he wrote, "Natural selection a true cause, modifying, but not sufficient cause to explain all phenomena. Climate a true cause but not sufficient" (quoted in Looker and Porteous 166). He seems to be seeking a more synthesized scheme that considers natural selection, species and population dynamics, and changing ecosystems—the triad of today's evolutionary ecology. Using the concepts of catastrophe, extinction, founder effects, succession, and pollution stress, Jefferies's literary fiction supports a narrative dynamic of the damaged world recovering toward something more whole and stable. Beginning with chaos and growing toward coherence, successional narrative can capture evolutionary ecology across many human generations. Jefferies felt that natural selection and climate were both true causes, but neither alone was sufficient to capture the complexities of emergent ecosystems. The fusion of the two concepts through narrative time captures this effect. Chaos provides, if not a tabula rasa, at least a simplified starting place at which to begin to model a recovering landscape.

The first section of the novel, titled "The Relapse into Barbarism," is the most innovative portion of what is otherwise a classical heroic romance. The unnamed narrator is a philosopher of science who recognizes that his version of natural history is one of several possible explanations of how nature has evolved to its postindustrial state. He approaches the history after the city of London came to an end by relying on relative perceptions involving various epistemologies, without appealing to an objective, scientific ideal of ultimate truth. Still, he is confident that mitigated realism is possible through the scientific methods of close observation and hypothesis that he employs to make sense of the past.

Like Gilbert White, Jefferies's narrator is a monographer whose task is to narrate a history based on a series of inscrutable ecological events. In this vision, as in a court of law, the truth resides with those who can articulate the most believable narrative. Rhetoricians show how ethos, pathos, and logos gain credit with audiences when used in synergy. Jefferies's narrator has a masterly eye for detail, and he brings the reader around to his version of things, not by claiming absolute correctness but by developing his ethos as a fully considerate, open-minded interpreter of the book of postmodern nature.

The narrator begins: "The old men say that their fathers told them that soon after the fields were left to themselves a change began to be visible. It became green everywhere in the first spring, after London ended" (11). Particularly compelling is the description of domesticated species, farm and companion animals, and agricultural plants either going extinct or returning to a feral state in the posthuman world. He gives a detailed tally of species that survived the cataclysm and have evolved into new niches, including several varieties of feral dog that now pose a threat to the remaining humans. With the plague as a working explanation for the mass death of industrial human populations, the narrator is left to speculate on the recovery of the landscape. A massive inland lake defines the geography of the new England, and the narrator insists that "the Lake itself tells us how it was formed," involving "changes of the sea level and the sand that was brought up [the Thames and] must have grown great banks, which obstructed the stream" (42–43). He entertains two major theories on the process that created this lacustrine environment: gradualism and catastrophism. His theory is gradualist, borrowing its aesthetic from the Neptunism of geologists Abraham Gottlob Werner and Charles Lyell. Through an accelerated gradualist process, the choked river Thames "began to overflow up into the deserted streets, and especially to fill the underground passages and drains, of which the number and extent was beyond all the power of words to describe. [. . .] [L]astly, the waters underneath burst in, this huge metropolis was soon overthrown" (43). Human infrastructure, when not maintained, falls into entropic decay under natural forces. After a period of thirty years, the narrator theorizes that the lake reached equilibrium with sea levels, and the extreme reaches of the lake to the east (at the Thames) and the west (at the Severn) came to exhibit daily tidal exchanges with the ocean.

In the narrator's natural history, the disequilibrium imposed by a saltation event achieves natural balance around a new landscape of open water, and the lake in *After London* becomes both the focal point for scientific explication and the playground for its hero, the fortunate Felix Aquila. By providing modern England with its own self-contained sea, Jefferies effectively recenters what's left of civilization around the shores of a new Mediterranean. England becomes an inward-looking little world of vying human bands and strange natural forces, more biologically heterogeneous than Victorian England had been. *After London* is a microcosm experiment that models the dynamics of a world made new by environmental upheaval. The novel's description of a flood-engulfed London partakes of the long history of speculation on London's susceptibility to rising oceans. London frequently soaked in flood tides in the medieval period, most famously when several arches of London Bridge collapsed in 1281 (Cracknell 89–102). Modern London has flood barriers that civil engineers see as entirely inadequate to hold off rising seas in the twenty-first century.

The narrator is a gradualist of the Neptune school, but he also introduces the theories of his rival, Silvester, who is a religious catastrophist: "those whose business is theology have pointed out that the wickedness of those times surpassed understanding, and that a change and sweeping away of the human evil that had accumulated was necessary, and was effected by supernatural means" (25). Silvester's chao-theistic scheme of divine retribution involves the astrological physics of "some attractive power exercised by the passage of an enormous dark body through space," which affected the earth's axial lean and "altered the flow of the magnetic currents, which, in an imperceptible manner, influence the minds of men" (25). In Silvester's reality, the psychology of catastrophe is material, involving magnetic attraction on the cognitive level.

This pseudoscientific way of explaining traditional Old Testament divine retribution is part of geological catastrophism's legacy to the scientific debates around the turn of the nineteenth century. Although the geologist Georges Cuvier had no interest in corroborating biblical accounts with his fossil-based theories of catastrophe, evidence of upheavals in deep time, like Noah's flood, tempted natural theologians to intentionally read the geological evidence as material support for the Bible's stories. When drained of its apocryphal theistic brain science, Silvester's cosmic catastrophism shares aesthetic ground with the Alvarez hypothesis that a meteor impact caused the Cretaceous extinctions that killed the dinosaurs and most other large organisms sixty-five million years ago. Chaos ecology makes much of the influence of random catastrophic events, and Silvester's "enormous dark body" falling from space is a meteor. Outside of Christian eschatology, scholars have suggested that ancient folklore and mythology depicting the wrath of planetary gods represents actual ancient human experiences with natural disaster, including meteor impacts (Palmer 206).

After London appreciates vying perceptions of gradualism and catastrophism and the various mechanisms that may cause ecological disturbance. The narrator teaches us to believe that the English were devastated by plague and that the infrastructure of Victorian civilization was left to the ravages of the floods. But he does not completely deny alternative interpretations. Like a good rhetorician, he simply places them in the light of ridicule. Since the narrative is subject to interpretation, Nature stands out as a mysterious, dynamic character, and stories about her develop into a new species of bildungsroman. The environment of postapocalyptic London flourishes by virtue of a complex character development, and the sensationalism of her uncertain past creates a new species of Gothic romantic heroine. With this dark, mysterious, compelling, moody, self-absorbed character of Nature, who is capable of self-improvement through evolutionary change, Jefferies has shifted the locus of the traditional nineteenth-century novel to the natural world. This new ecological heroine arises from the wreckage of a tragic past and moves toward fortuitous self-recovery by promoting species adaptations that fill niches in the new world.

The first part of the novel, "The Relapse into Barbarism," might more accurately be called "The Resurgence of Complexity." The "barbarism" of the new human culture is upstaged by the elaborate adaptations of fallow land. Victorian naturalists ardently sought ecological evidence of the self-restorative capacities of nature, as they provided a rationale for restoration projects that helped landscapes recover from industrial abuses. Although the narrator is too rational to fully anthropomorphize Nature as a mother, or to speak of an entity that is more than the sum of individual animal groups and flora left to lie fallow, Nature is implicitly presented as a character by the absence of other characters in the first part of the novel. Lots of life exists in the forms of animals and plants, but rather than existing independently, they are arranged and energized by the greater body of the land and waterscape.

In this opening section, Jefferies encounters linguistic challenges similar to those Darwin faced in attempting to narrate the story of evolution by natural selection. English requires an active agent to own the verbs in sentences, and this rule causes an unintended characterization in agentless processes. Darwin apologized in later editions of *On the Origin of Species* for personifying the word Nature, which he took to mean the action of natural laws. The recalcitrance of the active voice becomes clear in Darwin's writing: "It may metaphorically be said that natural selection is daily and hourly scrutinising, throughout the world, every variation, even the slightest; rejecting that which is bad, preserving and adding up all that is good; silently and insensibly working, whenever and wherever opportunity offers, at the improvement of each organic being in relation to its organic and inorganic conditions of life" (*Origin of Species* 66). *After London*'s author owes his readers no apologies for anthropomorphism, because Nature is his active character.

By collapsing the time required to observe evolution into only a few human generations—a shift from the evolution of species by natural selection to the evolution of a landscape by succession—Jefferies casts a clear beam on his evolutionary vision of nature's spontaneous movements from disorder toward a new order. Using the model of a self-contained England, Jefferies narrates the history of what could happen when disturbed land is largely left to itself, and when human civilization succumbs to intrinsic natural forces that work and technology have only temporarily kept tethered. It takes only one generation for most of the industrial human legacy, the culture of the ancients, to be swallowed up in the undergrowth.

Nature's succession follows a pattern of fortuitous opportunism among extant species. In the opening pages of the book, Jefferies treats the accumulation of time both cyclically and linearly: first, the four seasons of a single year wreak havoc on the formerly controlled agrarian countryside; meadows are "not mown" and the wheat fields have "no one to reap," opening that bounty to "clouds of sparrows, rooks, and pigeons [. . .] feasting at their pleasure" (11). Complex new interconnections arise by only the second year, when rapidly colonizing species occlude human paths and naturally reseeded fields grow a ghost crop of staples, which are again devoured and turned under by the onset of winter. Although the aesthetic discord of briars and brambles instills an image of unchecked and unproductive nature, that which "helped to destroy or take the place of the former sweet herbage," these stages of initial succession gradually give way to the secondary succession of slower-growing, more established species over generations (12). These are the figures of grandeur in a mature forest, the species that would form ecologist Frederic Clements's theory of the climax community in the early twentieth century.

Succession in this ecological narrative is both chaotic and coherent. Coherence is borne on an upward continuum, from the initially weedy, insubstantial undergrowth to a more stable complement of large trees.

Chaos is equally important, though perhaps less apparent. The plant species that happen to have seeds in the soil of disturbed land, that survive the grazing of whichever animals happen to remain, that happen to be distributed by whatever winds blow, storms disperse, or animals drop—these are contingent factors upon which the eventual stability of a community depends. From a distressing vision of a harlot Nature "starting from all sides at once" to extirpate the human legacy, time changes Nature into a mature, comely forest ecosystem: "protected by the briars and thorns from grazing animals, the suckers of elm trees rose and flourished. Sapling ashes, oaks, sycamores, and horse-chestnuts, lifted their heads. Of old time the cattle would have eaten off the seed leaves with the grass so soon as they were out of the ground, but now most of the acorns that were dropped by birds, and the keys that were wafted by the wind, twirling as they floated, took root and grew into trees. [. . .] [T]he young trees had converted most parts of the country into an immense forest" (12). From keys to trees goes this succession, from the vines and brambles up to the grand species that are metonyms for England itself, the elms and oaks. The progress from humans' artificially sterile farm fields to nature's complex ecosystem, the mature forest, is enabled by an intermediary imbroglio.

Ecology's contrasting aesthetic of chaos and coherence can be reconciled when landscapes are considered over time: ecosystems undergo directional change after disturbances and then settle into a more stable cyclical condition, until the next disturbance lurches the system into a new disequilibrium. Anticipating later ecological philosophy, Jefferies's detailed description of natural succession introduces humanity as a force of ecological disturbance whose impacts can nonetheless be remediated quickly. The legacy of the industrial ancients dissolves as the species community shifts toward its next stage of complex organization. Jefferies is an optimist about the recoverability of Nature.

Although the established forest could be viewed as a predetermined telos to which nature inevitably returns when left to itself, there is a subtle but necessary distinction between the greater balance and grandeur of any mature natural system and the theory of a superorganism climax community with a particular species composition that has synergy and balance absent in earlier stages of succession. The latter carries a sense of design and teleology that has not been borne out in ecology's recent experiments. Disturbances happen on many different scales, and there is a constant background of small disturbances, such as a tree falling and opening up a sunny area in the canopy, or a beaver damming one stream rather than another. Chaos is not just mind-boggling ecological revolution; it is also, perhaps more significantly, the tiny ripples of transition that follow every fortuitous change in a landscape.

Once the novel turns to Felix's adventures on the lake, the site of ancient London becomes his ultimate challenge. It is a submerged dump resembling a Superfund site. Its toxic quality makes it the longest-lingering legacy of the ancients in the new primordial world. The novel's greatest sensation comes from the ecological impacts of the collapse of industrial society. Felix is warned of the former city's proximity by a flock of fleeing birds, and he is fortunate to enter the miasma when winds have dispersed the toxic fogs. Poisons seep up from the submerged city and adorn the landscape in a hellish menagerie of fire, killer fog, and slimy soot that shellacs the skin of any explorer who pursues the lost treasures of the ancient civilization. This hellscape gives concrete form to Jefferies's moral admonition against a consumerist industrial society and its dystopian legacy. It is a sensational scene, the neosublime of Jefferies's postapocalyptic vision, an antithesis to the rest of the novel's purity. As the only manifestation of pollution that the surviving humans know, the deep, nasty muck of old London is a vision of the closed circle of human activity and environmental pollution. Animals exposed to contaminated environments pick up a body burden of toxic compounds. Pollution can never just be cast away; it finds its way back into the ecological system that supports life on a closed-system planet.

These graveyards of pollution bring to the surface an interesting question that ecological writers like Alan Weisman have put into narrative form: what happens to our industrial infrastructure if no one is around to maintain it? Even if industrial practices are not directly responsible for some catastrophe-to-be, how would the neglect of our machines create new problems? Particularly problematic are two inventions unknown to Jefferies's era: radioactive waste and plastic (Weisman 114–18). Narratives based on scientific predictions have attempted to measure the footprint of the industrial world through deep future time. We will return to consider twenty-first-century versions of Jefferies's narrative in the next section.

After London develops a complex relationship between seeming opposites: degraded industry and wild Nature. Jefferies's interbred vision permits industrial-ecological futurity to have a life beyond the extinction of the Victorian realm. Cities and industry have succumbed to disaster, but their deep-seeded ecological effects live on as the new, primitive world's purgatory, the site of intense toxic corruption that is also the gateway to impossible wealth. By secularizing religious rites and positioning a quest narrative on the stage of a postindustrial environment, Jefferies recasts Romantic natural super-naturalism in a modern, deep-ecological mold. Felix overcomes the nasty inheritance of his extinguished elders and founds his new green world on the principles of sustainability, community, and harmony. He retrieves the jewels of the ancients from the jowls of London, muses on his luck and personal fortitude, and rises to apotheosis among his new brethren of the shepherds, who are the gentle folk among the more violent tribes in the new world.

His success comes from an ability to pay close attention to Nature's patterns: the tides and winds that direct his canoe around the lake, the exodus of birds that warn of the submerged city's proximity, the health of landscape heterogeneity, the instinct of dwelling stimulated by shepherd land. Jefferies's final images are overtly idealistic, with a back-to-the-earth narrative ending in the romantic comedy of Felix bringing his future bride, Aurora, to her new home. The comedic ecological mode releases the tension of London's corruption, but it does not erase the poignant binary of regional utopia and dystopia revealed by the novel's journey.

Within the same 1880s zeitgeist, John Ruskin delivered his "Storm-Cloud of the Nineteenth Century" lecture to the London Institution in 1884, one year before After London arrived in bookshops. Ruskin gives an eyewitness account of the environmental effects of industry just as coal burning was intensifying in British urban centers. The essay reveals the psychological and physical stresses wrought by chaotic weather that Ruskin calls the "plague wind" of his industrial age. Just as Ruskin expressed nostalgia for craftsmen in The Stones of Venice, he also looks back wistfully to a time before widespread coal pollution. Jefferies, by contrast, imagines the natural future after industrial pollution.

The eruption of Krakatoa in 1883 contributed to the atmospheric chaos of the era. This eruption had apocalyptic qualities similar to those of Laki, exactly a century earlier, and those of Tambora sixty-eight years before. Northern Europeans experienced abnormal cold from 1883 through 1888, resultant crop failures due to heavy rains, and major optical effects, most notably, spectacular sunsets until February 1884. Painters had a bonanza with the surreal colors of the ash-laden skies over Europe and America: it was not only the vivid beauty that captured their attention but also the anxieties of the modern age, as evidenced in Edvard Munch's The Scream, with its psychedelic sky (Olson, Doescher, and Olson 29). Krakatoa's ash cloud led to the scientific discovery of the jet stream, which observers originally called the "equatorial smoke stream," the global weather maker (Winchester 288). Krakatoa's eruption was one of the first globally experienced environmental events in the new age of telegraphs, so, unlike Laki and Tambora, Europeans knew about the eruption within hours and were able to monitor its effects on their own skies and coastlines. Krakatoa's boom sent a periodic air concussion around the world that could be measured in barometers and barographs at two-hour intervals, and the tsunami originating in Java made it all the way up to southern France as seven distinct three-inch ripples (Winchester 278).

This collective global experience and the ability to communicate about it rapidly allowed remote cultures to recognize physical interconnections in the macrocosm (Winchester 269). Environmental events could no longer be viewed as isolated and regional; they potentially had global effects on ecology, climate, epidemiology, and human health. Jefferies, a man constantly wandering and wondering in the English countryside, may have thought that the bizarre weather caused by industry and Krakatoa's eruption was a sign of an ominously changing future. *After London* is a wish for chaos to make the world over. Although he did not live to see the end of Krakatoa's climatological effects in 1888, Jefferies wrote in 1887, the year of his death, "I look at the sunshine and feel that there is no contracted order: there is divine chaos, and, in it, limitless hope and possibilities" (*Old House at Coate* 163).

Jefferies's novel delivers an implicit message that ecological crisis in the industrial era is not solely a crisis of nature but also a crisis of civilization. Natural upheavals mirror the violent ways in which industrial society treats its environment. In Shelley's *The Last Man*, violence is seated in colonial exploitation, where the British appropriation of natural resources from lush and spicy climates also involves the import of tropical sickness, a byproduct of imperialism that caused anxiety during Britain's colonial apogee. With waves of consumer goods coursing through British ports in the 1820s, Shelley's novel depicts the temperate English climate washed over by tropical breezes, and so the estrangement and psychological stress extend beyond epidemiological paranoia into the global mentality of the modern age of climate change. *After London* demonstrates, sixty years further into the Industrial Revolution, in the 1880s, how there is no longer any gap between natural and anthropogenic catastrophe. There is no possibility of "natural disaster" in its original sense, only the more sinister pattern of cultural by-products terrifically elevating baseline levels of ecological disturbance. Landscapes absorb the cocktails of industrial air and water pollution and disperse these toxins back out into the living bodies they support.

The question of nature versus nurture becomes a dialectic of influence that can now apply to anything in nature—not just animals but landscapes themselves. Only the demise of that perpetrating culture will allow the character Nature to step back into an active role of self-recovery, which Jefferies imagines, with considerable theoretical success, as a process of flooding and ecological succession.

The Time Machine

The role of human accountability in ecological degradation during the hundred-year span between Gilbert White and Richard Jefferies only intensifies as we look at the social anxieties and evolutionary intrigues of the fin de siècle. H. G. Wells was an ardent student of evolution and the new science of ecology. He recognized that the ecological succession described in *After London* was essentially "the same competition" as evolution by natural selection, only cast in different scales of time (Wells, Huxley, and Wells 989). In his late-career 1929 collaborative work with Julian Huxley, *The Science of Life*, Wells looks to congruent scales in nature, both physical and temporal, to enrich the study of evolutionary ecology.

> There is another way in which the little mirrors the big. The same competition which results in the comparatively speedy development of ecological succession results also in the portentous slow development of evolutionary succession. A landslip or man's destructive hand uncovers a patch of the bare earth, or impounds a body of barren fluid; it is colonized by a succession of communities, and in a few decades is tenanted with rich life again. The whole world, both land and sea, was once free of life; and aeons later all the land was still one great bare patch of earth and rock. First the seas, and then the lands were colonized. In both there has been a succession of faunas and floras, each one on the whole exploiting the environment a little more effectively than the one before. Evolution is a slow succession of a series of ever new and ever improved communities towards a still unrealized climax. The most up-to-date of life's existing communities are still very wasteful in their exploitation of the world's resources. It

remains to be seen whether man, with his deliberate aim at a higher efficiency, his replacement of the hitherto dominant tree by his own cultivations and devices, will make a mess of things and fail, or will succeed and hold on from climax to climax. If he fails the forest will return. (989)

Jefferies predicted the failure of human devices and the return of the forest. The mature H. G. Wells was undecided, but still enchanted by the questions of future human evolution that he raised in his youthful novel *The Time Machine*. He was clearly influenced by contemporary ecological discourse, finding Frederic Clements's terms "succession" and "climax" useful polemics to counter a more chaotic and contingent view of evolutionary change in nature. This older Wells saw evolution as inherently tending toward "improved communities," greater complexity, and higher human potentials for advanced thought, technology, and perhaps even civic harmony.

Almost forty years earlier, Wells asked the same question of whether "man, with his deliberate aim at a higher efficiency [. . .] will make a mess of things and fail, or will succeed and hold on." But the young Wells of 1895 clearly had no faith in the climax of higher civilizations over time. *The Time Machine* is an enduring novel precisely because of its sardonic cynicism about the prospect of improvement, given the initial conditions of Victorian industrial society. By "bursting the limits of time," to use the geologist Cuvier's phrase, Wells envisions bizarre posthuman forms in the remote future whose relationship to contemporary humans is analogous to the uncanny resemblance between humans and our fossilized ancestors. This is the essence of Wells's chaos: his radical vision of evolution as indeterminate and emergent downstream of initial conditions, rather than proudly, inevitably progressive, as Darwin saw it. Wells's Morlocks are to the future what *Australopithecus afarensis* is to the past, and both are "lesser" than *Homo sapiens* when considered culturally rather than ecologically. Victorian science had a tremendous bias, inherited from religion and philosophy, that placed humans above other animals as superior in intellect and civilization, even when these capacities got our species into ecological trouble. This era in evolutionary biology passed on the progressive, hierarchical bias to ecological science, with its early paradigms of succession and climax.

The younger Wells would not accept such Victorian progressivism. *The Time Machine* lurches forward eight hundred thousand years, a narrative move that is chaotic in the modern sense because initial socioindustrial conditions in 1895 are determinative through evolutionary time. From the industry that divided people into the capitalist working and ruling classes, and from the machines that made this division profitable, Wells creates a narrative in which the emergent social properties of occupation, spatial division, diet, and disposition become the prevailing topoi that impose biological changes in the future. In this industrial society that unfolds over nearly a million years toward the Morlock state, the conditions that Wells saw as most characteristic of his time have grown from seedling social shifts into full-blown physiological imperatives wrought by evolutionary selection. No narrative of Gilbert White's or Mary Shelley's pre-Darwinian era could have achieved Wells's bizarre vision of dystopia; *The Time Machine* takes biological evolution and the conditions of industrial modernity as launchpads, and he propels these ideas through the exacerbating eons.

The Time Traveler's series of insights on the nature of the future world shows his personal evolution from Darwinian idealism to the chilling evolutionary pragmatism that best explains his observations of the Eloi-Morlock dynamic. Whereas at first he views the world as a warm garden, its people delighting in their labor-free existence (with the minor liability of intellectual death), he comes to understand that the Eloi are "mere fatted cattle" serving the appetites of the more intelligent Morlocks, who live in an industrial underworld (62). This is no Enlightenment dream of perfectly regulated nature and well-oiled socialism; it is a sinister vision of degradation set on industrial rails through deep time. The stark evolutionary disjuncture between the Traveler's Victorian England and the Morlock world instructs us about manifest social, behavioral, and physiological differences wrought by evolution, and these contrasts only highlight the features of Victorian England that remain: its social inequality and its obsession with machines.

The power of deep time is charged by the Time Traveler's sense of ultimate isolation when his machine is stolen by the Morlocks eight hundred thousand years into the future. Our brains struggle to imagine the passing of thirty-two thousand generations, each hungry in their own time. For any coherence, the narrative needs to leap into the future, merely flashing on spots of time that provide evidence of how changes occurred along the way. Hyperevolution results from rapidly changing environmental conditions, and these punctuation events are taken as intensive periods of change in a chaotic environment. Rather than imagine a cataclysm that causes hyperevolution, as Jefferies did, Wells puts the reader in hyperdrive through eight hundred thousand years to gain an evolutionary reference point on 1895. The novel's images of time travel are marked by the cosmic rhythms of the sun and moon, the ecological rhythms of seasonal change, and the diurnal flickering that accelerates toward a blended gray of simultaneous night and day. The only trend that breaks from natural cycles during the time warp occurs when the Time Traveler observes "a richer green flow up the hillside, and remain there without any wintery intermission" (19–20). The earth has become warmer, and to the utopia-minded traveler, a richly verdant terrain promises a more advanced civilization.

Utopianism descends the staircase one false hypothesis at a time until it reaches the dystopian depths, the machines in the basement. The Victorian spatial discrepancy that entombs the workers in dark factories and leaves the ruling classes in the leisure of open air is rendered into a higher-order evolutionary trend, a wedge driven between social classes. Allopatric speciation, which occurs when two populations of a single species are spatially separated and evolve apart as they adapt to different conditions, is the evident mechanism that split one species of *Homo sapiens* into two, the Eloi and the Morlock. The state of the environment in the distant future, a "long neglected and yet weedless garden," seems to follow the ideal of progression toward a "climax. One triumph of united humanity over Nature had followed another" (26, 31). But the idyll takes an ironic turn: the Edenic surface in 802,701 is a veil over an insidious engine-driven reality. The machine remains central, but it is now the machine under the garden, and Eden is a technosphere, a mechanic's paradise. Rather than project an environment overtly destroyed by industrial pollution and urban sprawl, as so many futuristic dystopias do, Wells is more nuanced in having his Time Traveler vacillate between scientific hypotheses of utopia and dystopia. The Traveler's initial vision of an ecological utopia is based on aesthetics. His observance of the Eloi's labor-free, fruit-eating existence temporarily blocks his perspective on where they might fit in the trophic web of this ecosystem. He assumes that they are the top frugivores and that they long ago eliminated any threatening predatory species, as humans tend to do.

The realization of dystopian nature comes with the insight that this world is an energy-exchanging system in which one species systematically preys on the other. The static-state aesthetic utopia of Eden falls apart with his subsequent and stronger theory of systems ecology, which allows him to trace the flow of energy and nutrients throughout the system. From a utopian telos that imagines the Eloi state "settled down into perfect harmony with the conditions under which it lived [. . .] the last great peace," he eventually concludes that there is no resting place for evolution (33). Correlatively, the environment has no static state; both processes are involved in ceaseless turmoil, moving from one state to another and subject to chaotic fluctuation. The Morlocks are only temporarily in control of the system. The Time Traveler's launch into the remotest future shows a senescent Earth ruled by crabs and butterflies.

The climate is hotter in 802,701 than it had been in 1895. Although he avoids a singular explanation for global warming, the newly arrived Traveler promotes contingency:

> I think I have said how much hotter than our own was the weather of this Golden Age. I cannot account for it. It may be that the sun was hotter, or the earth nearer the sun. It is usual to assume that the sun will go on cooling steadily in the future. But people, unfamiliar with such speculations as those of the younger Darwin, forget that the planets must ultimately fall back one by one into the parent body. As these catastrophes occur, the sun will blaze with renewed energy; and it may be that some inner planet had suffered this fate. Whatever the reason, the fact remains that the sun was very much hotter than we know it. (44–45)

Perhaps combustion of the planet Mercury, or Venus, can explain the intensified heat of the sun? As in the earlier narratives, the Traveler advances no industry-based theory on warming, but this speculation comes before he is aware of any industry at all in the pastoral new world. The Eloi enjoy a half-clad existence almost wholly outdoors, and the inconveniences of seasonal variation are unknown. The very absence of environmental challenge contributes to the Eloi's cognitive oblivion. The place that had been Greater London now blooms as a strange tropical paradise: "You who have never seen the like can scarcely imagine what delicate and wonderful flowers countless years of culture had created. [. . .] My general impression of the world I saw over their heads was of a tangled waste of beautiful bushes and flowers, a long-neglected and yet weedless garden" (25–26). This depiction of a pleasant nature cloaking a terrifying new world order hearkens back to Mary Shelley's tropical England. The language is intensely descriptive because of its paradoxes: the beautiful waste, the weedless neglect. Cultural selection imposes artificially bred beauty, and the climate acquiesces under technological force. The flowers become the symbol of this future world because they are the only piece of evidence that the Traveler brings back to his own time. Their fruit is consumed by the vegetarian Eloi, and herein lies another paradox: the flowers are the blooms of machines. Although their principle of growth may be organic, using water, air, and sunlight to make carbon-based matter, their existence is owed to the technology that created them. Like a biotech crop that uses borrowed genes, the better to adapt to environmental extremes, these flowers exist because they have been genetically manipulated by their humanesque creators. The Eloi, the Traveler ruefully comes to acknowledge, live on a glorified feedlot controlled by Morlock technology. Their tender flesh and underdeveloped minds make them perfect for domestication, and one posthuman species systematically breeds and devours the other. This is not a lobby for vegetarianism; it is a vision of a world grown so technological that every level of what used to be nature is now regulated as part of a thermodynamic system, a mechanization of systems ecology. Thermodynamic systems don't have morals; and it seems that humanity's heirs, the Morlocks, have no qualms about eating their siblings in ancestry, the Eloi. It is something like a human eating a chimpanzee, which our systems make taboo and generally illegal.

Accompanying this amusingly dismal insight into the posthuman predator-prey dynamic is one further blow to evolutionary teleology. The Traveler discovers the decay of all knowledge as a consequence of intellect's obsolescence. Moral systems are born of human intellectual capacity. One way that humans through the ages have divided themselves from other animals (perhaps speciously) is by a perception of our unique capacity to do the right thing, sometimes in spite of self-interest. Altruism is the subject of intense debate in evolutionary biology. Although the Traveler endures a moral blow by observing the Eloi's indifference to Weena's near drowning, a deeper despond results from the discovery of the decayed Palace of Green Porcelain, a proxy for Victorian England's Crystal Palace. Leafing through the disintegrating books that were once part of a massive library, the Traveler says, "Had I been a literary man I might, perhaps, have moralized upon the futility of all ambition. But as it was, the thing that struck me with keenest force was the enormous waste of labour to which this somber wilderness of rotting paper testified. At the time I will confess that I thought chiefly of the *Philosophical Transactions* and my own seventeen papers upon physical optics" (67–68). A literary man, he suggests, would brood over the philosophical "futility of all ambition," whereas the scientific mind turns to the thermodynamic, entropic concerns surrounding a "waste of labour." But the sense of regret is the same. *Philosophical Transactions*, the journal of the Royal Society of London, was the most recognizable scientific authority in the nineteenth century, but the Traveler's contributions literally turn to dust when the social institutions that cradled his work are pulled under.

The remnants of a decayed civilization form the new wilderness. The future of nature is a technosphere, a mechanically regulated system; the future of educated culture is wilderness, where there are no markers of epistemological history, paradigms, or creative movements. Those who wander there are lost without reference points. This crossing over between culture and nature suggests that the future of nature is to be acculturated, mechanized, modified, regulated, polluted, cleaned up. Learned knowledge suffers a similar fate: if it is economically useless, like the wastes that the Scottish highlands or Floridian swamps used to be figured as, no energy need be expended to maintain it. Ecological entropy is kept at bay by the machines; intellectual entropy advances. Popular culture has started to take stock of the degradation of intellect in a culture of hypertechnology, cheap sensationalism, and physical gratification, as we see depicted in the 2006 satirical science fiction film *Idiocracy*. Like the Morlocks, we increasingly spend our time inside, in machine-controlled climates tending machines.

The Traveler's final great leap into the deepest future completes Wells's vision of eternal ecological drift toward no particular endpoint under the mad scalpel of entropy. Several million years further into the algorithm of evolutionary selection, the Traveler finds his alienation from the surrounding life complete: "I cannot convey the sense of abominable desolation that hung over the world. The red eastern sky, the northward blackness, the salt Dead Sea, the stony beach crawling with these foul, slow-stirring monsters, the uniform poisonous-looking green of the lichenous plants, the thin air that hurt one's lungs; all contributed to the appalling effect. I moved on a hundred years, and there was the same red sun—a little larger, a little duller" (83). The earth has lost its axial spin, leaving its odd inhabitants in perpetual, chilled twilight where the city of London used to be. The Thames Valley, which had burst with sinister vitality under the Morlocks' cultivation, has long aged and now invokes the poisonous sublime where its picturesque garden had once grown for the Eloi. From his uncanny similarity with the posthumans of 802,701, the Traveler may as well be an astronaut visiting an alien planet; from a paragon among the Eloi, he has devolved into prey for enormous crustaceans. Even though the man remains the same, his position in the environment of this extreme future time is relative to its condition, not his; he is further displaced in evolutionary time. This final vision of utter desolation, which presumes that all human descendants are extinct, cinches Wells's thesis that the narratives of the deep future are fragmented, only partially coherent, and completely subject to chaotic chances imposed on life.

In Wells's novel, the Romantic quest of an inspired man pushing the limits of knowledge undergoes a crucial revision for the audiences of late Victorian England. The Traveler barely survives to bring his story back to the comfortable lounge where his audience is assembled to judge the bizarre tale. This staged opposition of armchair philosophy with sinewy enterprise piques the concerns of a positivist era. How can hypotheses be tested when evidence is contained in superhuman time frames? Can scientific epistemology disentangle truth from hope to arrive at a valid evolutionary theory? The Traveler ends his tale with a philosophical disclaimer that brings his outrageous claims into the controlled sphere of a gentlemen's supper: "I cannot expect you to believe it. Take it as a lie—or a prophecy. Say I dreamed it in the workshop. Consider I have been speculating upon the destinies of our race until I have hatched this fiction. Treat my assertion of its truth as a mere stroke of art to enhance its interest. And taking it as a story, what do you think of it?" (87). Involved as he is in a metadiscourse with the sciences of his tutelage, Wells has used literary narrative to advance a creative argument about the nature of scientific prediction, the patterns of evolution, and the legacy of Victorian industrialism.

A prophecy is the same as a lie; both might come true. From the wide terrain of possibility spring a few fortuitous seeds. Once the future has happened, we can trace its origins in the past and pretend that time connotes destiny. But this is a lie that substitutes necessary causality and design, the telos, for the mere necessity that *something* succeeds among the many forces vying for success. Progressivism is hacked apart by this farce of evolutionary mischance and the further vision of a moribund earth in the remotest future. The assembled gentlemen dismiss the account as just a story, excepting only the withered alien blooms the Traveler has recovered from 802,701. This little piece of evidence abuts narrative chicanery with a physical artifact, effectively suspending judgment indefinitely. The Traveler brazenly disappears into another time once again, leaving his fate and further discoveries as unknown as the veracity of his first journey. The novel ends, "as everybody knows now, he has never returned" (90).

An ecological reading of these two Victorian novels reveals the role of chaos and disturbance in industrial nature's narratives. With advances in atmospheric science and evolutionary theory, Jefferies and Wells were able to create bizarre new ecological realities as the legacies of an industrial world order. Both texts render a more incisive narrative about ecological disturbance in the modern era of secondary nature than we find in reactionary wilderness texts that call for environmental preservation. While all four of these works could accurately be called environmental because of their implied ethics, fruitful readings result from considering them as works of early ecology. Chaos elicits vital questions in the sciences of the environment, such as the nature of ecological disturbance and recovery. The narrative fragmentation that characterizes each of these four works reminds us of repeating conceptual failures in linearity and periodicity when it comes to predicting the dynamics of disturbed ecosystems.

As theorists of the twenty-first century are learning, an accurate narrative vision of the coming centuries and climate change requires our acceptance of chaos as a player in future scenarios. With initial conditions, random disturbance, emergent relationships among components, nonlinear positive feedback loops, and tipping points to consider, there are plenty of qualitative and aesthetic bases for accurate literary descriptions of modern ecological processes. Where Gilbert White speculated on the downstream effects of aberrant weather and volcanic eruptions, climate scientists now use data from eruptions to predict climate anomalies and imagine radical correctives to global warming, such as the infusion of sulfur dioxide into the earth's atmosphere. Where Mary Shelley conceived a tragic drama out of colonial epidemiology and global warming, the CDC in America monitors patterns of disease in high-density urban populations and attempts to calculate how a warmer climate will affect the spread of disease. Where Richard Jefferies imagined the long-standing toxic legacy of industrial London, environmental agencies now calculate the pervasiveness and persistence of industrial and agricultural chemicals in the environment and use these assessments in regulation. Where Wells envisioned a garden cultivated by machinery, industrial agriculture has grown into a petrochemical machine-driven system, and biosphere projects attempt to create mechanized ecosystems in microcosm. By working through the anxieties of nineteenth-century British culture, by infusing scientific ideas with a spark of literary imagination, and by situating evidence within the architecture of literary narrative, these works anticipate scientific theories of the modern industrial state of nature.

3

With these Romantic and Victorian perspectives in mind, we may now consider how chaos has evolved into a theoretical paradigm in today's ecological sciences. It is surprising that narrative has been so little used in more than a century of scientific ecology. After all, the most important work of ecology is Darwin's *On the Origin of Species*, which uses narrative as a medium despite the challenge of articulating a plot that accommodates deep time and implies agency in what is truly a blind, agentless process. Today's science students are not often assigned classic narratives of ecology; even the foundational ones by Darwin and Wallace are not part of the standard science curriculum. It may not be that narrative is inherently antiscientific, as our culture seems to assume by using the words "story," "tale," or "fiction," but that ecology has drifted away from the narrative practice with which it originated in order to coddle models.

Since physical modeling permits precise empirical controls, manipulation, and repetition, it raises the prestige of ecology as a "hard science" aligned with microbiology and chemistry. Narrative, by contrast, leans more toward the "soft science" of anthropology and, even further, into the humanities. Normal ecology's resistance to narrative can be explained at least partially by a disciplinary cultural resistance to antiscientific theoretical critiques from the humanities. Still, scientifically literate writers wield great power to manage the complexities and contradictions of whole ecosystem dynamics, while modeling in the fish tank has little power to negotiate among different levels of the system. A microcosm of a rainforest is little more than an abstraction, a vignette, a sketch simulacrum that simplifies and idealizes the original, thereby continually misrepresenting it. A narrative case history of the forest, on the other hand, can bear witness to a history of change and stasis, elucidate its dynamics based on observation in situ, and provide informed predictions about its future. Popular works by the likes of Wendell Berry, Michael Pollan, and Barbara Kingsolver, whose tales tell the fortunes of ecosystems, make the most of narrative ecology.

In literary studies, narrative is an enormous field of inquiry, and its most basic tenets question how meaning is created by telling stories. Narrative is fundamental to our cognitive development, effectively weaving a series of discrete events into the coherent articulation of individual experience that constitutes self-identity, acculturation, and identification with nature, or dwelling in a home ground. Reflective, intentional, and constructionist theories of narrative suggest different roles that narrative can play in relation to truth, from pure relation (reflective), to authorial designation (intentional), to the very creation (or construction) of meaning from a meaningless world. All three levels are of concern to the philosophy of science. Ideally, but perhaps never in fact, do science stories reflect objective truth. Very often, as with Darwin's grand gradual march of adaptation, the interpretation of data is skewed by culture and intentionality. Sometimes, as with static climax communities in ecosystems, the whole narrative may eventually be revealed as a fantasy of coherence that denies reality's chaos.

Ecological science is grounded in evolutionary theory, which naturally lends itself to narrative because the fourth dimension of time is the essential agent of change, though it is nearly imponderable owing to deep time's superhuman scale. Gillian Beer's literary study of Darwin complicates the ideal of pure reflection of truth by showing how science is constructed out of particular cultural contexts. Darwin's ideas of evolution depended on a slow, constant tempo of natural selection working on variant forms; this progressive gradualism may well have been a construct of Victorian society, which used a belief in improvement or progress to justify imperialism. This study questions gradualist progressive norms by examining nineteenth-century fictional narratives of environmental chaos based on natural disasters. Narrative is often an embedded engine that lurks beneath accounts of natural history written by scientists who are more overtly interested in facts than in stories, because the latter can be likened to myths that weaken scientific claims. Peter Bowler discusses the tension in evolutionary science

surrounding narratives of human phylogeny that "resemble folktales or creation myths, a suggestion which horrified modern paleo-anthropologists
who thought that it implied that they too were still only 'telling stories.' In
fact, all explanations of particular events in phylogeny which invoke adaptation have a narrative structure (often called an adaptive scenario)" (*Evolution*
282). When using narrative structure, there will always be a debate as to
whether the narrative purely reflects an external truth or is a construct of a
particular authorial ego or cultural condition. Ecological science often avoids
narrative so as to appear legitimate; after all, peer reviewers refer to the traditional standard of empirical repeatability and statistical support.

H. G. Wells, a man well aware of the rhetorical power of stories, wrote a
post-Darwinian scientific account of evolution in the early twentieth century. Wells is one of the few thinkers who straddled science and literature
well into the 1900s, and his book *The Science of Life* (1929) narrates modern
understandings of the life sciences without depending on fictional scenes to
illustrate them.[1] Wells's scientific account of evolutionary history weaves
together the facts of evolution with threads of modernist ideology, which
makes the work a subjective scientific narrative, just as Beer revealed Darwin's to be. An objective description of observed processes in nature is
continually elusive. Whereas the Victorians were enamored with Darwin's
coherent articulation of complexity, Wells's modernist narrative denies
anything resembling teleology or purpose in evolution:

> Variation is at random; selection sifts and guides it, as nearly as
> possible into the direction prescribed by the particular conditions of
> environment. Once we realize this, we must give up any idea that evo-
> lution is purposeful. It is full of apparent purpose; but this is apparent
> only, it is not real purpose. It is the result of purposeless and random
> variation sifted by purposeless and automatic selection.
>
> The term purpose has a very definite meaning. It is a psychological
> term, describing a certain familiar state of our own consciousness: it
> implies the prevision of an end, and a determination to reach that
> end. (641)

---

1. The book was written with co-authors Julian Huxley (grandson of Thomas Huxley and
brother of Aldous) and H. G.'s own son, G. P. Wells. Huxley, who is celebrated as one of the archi-
tects of the modern synthesis in biology, helped to lend this encyclopedic book credibility in scientific circles.

A sense of higher purpose is exactly what Darwin cultivated with his descriptions of beauty and adaptation. Darwin's Victorian aesthetics may have
grown out the desire to suppress public horror at the idea of an evolved, rather
than created, biosphere. In the passage quoted above, however, we find evolutionary narrative intriguingly updated for Wells's new audience. Although
purpose is useful in helping us conceptualize how comparative order has
arisen from comparative disorder through evolutionary time, Wells et al.
banish purpose from explanations of physiological change and assign the
appearance of purpose to human brains seeking design and pattern. Further,
the authors do not deny the overwhelming task of conceptualizing this
"drama," or the very small and recent part we have been given on the ecological stage: "Evolution is the sum of a swarm of processes, now independent,
now mutually interfering. The plot of the drama is not a single thread but a
tangled skein of hundreds of threads of which our own is only one" (786).
They chastise the antiquated view of purely gradualist evolution by gesturing
to the mass-extinction events that Gould and Eldredge would use in the 1970s
to make the case for evolution by punctuated equilibrium. In the story of the
Cretaceous extinctions sixty-five million years ago,

> the pressure of environment on life, a pressure quite external and for-
> tuitous, makes itself felt. The great climactic revolution that killed off
> the dominant reptiles opened the door of opportunity to the mam-
> mals: their warm blood enabled them to withstand the cold, their
> very smallness and insignificance was now a help when climate cut
> down the world's vegetable food supply. [. . .] [F]inally climate comes
> in again to extinguish many of the strange and exciting creatures which
> the same blind agency, by removing their competitors, earlier started
> on their evolutionary career. Change of climate may cause extinction
> directly, as it did with so many of the larger herbivores during the Ice
> Age, or indirectly. [. . .] Looked at thus, Evolution would seem to be a
> chaotic affair, its changes dictated by one accident after another, each
> one the outcome of the chance advantage of the geological moment.
> (788–89)

By making these bold and unflinching statements of historical relativity,
Wells et al. begin the important work of introducing chaos into the modern
ecological narrative. It is much trickier to narrate chaos than it is to plot
coherence. Passages like the one above are baldly opposed to the designed,

balanced, economical aesthetics of previous centuries, and the authors take care to explain how our cognitive processes formerly created such illusions in the guise of indomitable truths. Though Wells is attempting an objective description of chaotic biological processes, he is still indebted to narrative devices of order and causation in this history of evolution, even when that order arises from random events.

The ecological historian Donald Worster discusses the shift in ecology that promoted chaos as an offspring of computer-based mathematical models of the 1960s, which had found chaos in simple algorithms. Ecologists began to look with experimental scrutiny on the classic concepts of succession, climax, and equilibrium advanced especially by Fredric Clements and Eugene Odum, and these later scientists found a very different dynamic operating in nature (Worster 9). Replacing the older aesthetic of the mature climax ecosystem that is always reached when undisturbed, a vision emerged of a continually shifting mosaic of species and communities that never reaches a point of stability because disturbances, small and large, are continual. In the analysis of Drury and Nisbet (1973), ecological succession was merely the observation of "differential growth, differential survival, and perhaps differential dispersal of species adapted to grow at different points on stress gradients" (quoted in Worster 9). This conceptual shift permitted the discovery of ecological chaos, wherein nature is seen as inherently erratic, contingent, and very difficult to model for predictive purposes. If success in science is the ability to predict outcomes, chaos is a fundamental rift in the fabric of ecological modeling that has driven many ecologists inside, into computer-based virtual spheres where such complex behavior is somewhat more approachable (Worster 13). This revolution in thinking about nature as chaotic, Worster contends, rivals the conceptual revolutions of quantum mechanics and relativity that marked the beginning of the twentieth century, and it signals a further break from the Newtonian worldview of perfect balance and impeccable calculability (14). The modern paradigm of disturbance shows how a literary interpretation of disturbance in the nineteenth century was prescient about concepts that ecology now sees as fundamental.

Eugene and Howard Odum, two prominent microcosm ecologists of the twentieth century, took their study of the ultraliminal estuary ecosystem into the theoretical realm by including chaos theory in the dynamics of general complexity. Nature is fraught with edges, cusps, and cliffs. Most organisms make their niche in liminal spaces where elemental exchange is maximized, such as at the edges of bodies of water (rather than in the open ocean) and in terrestrial ecosystems. This reliance on interfaces demonstrates how the minimum requirements for life—water and an energy source—are clustered in specific places on the globe; location value heightens interspecies competition and energizes evolution. With the idea that "pulsing" is a pattern especially apparent in liminal locations, or ecotones, the Odums claim that random variations are intrinsic to the behavior of any complex system. The Odums' pulsing paradigm replaces balance with a narrative of chaos:

> The concepts through which we view nature are sometimes called paradigms. One of our comfortable concepts of nature visualizes growth followed by a leveling. In these days when society is beginning to recognize the limits of the biosphere, people, scientists, and governments talk of sustainability, that is, managing growth so that the life-support carrying capacity of the earth is not exceeded. The steady state is seen as a goal for such efforts as well as the final result of self-organization in nature. However, there may be a more realistic concept, that nature pulses even after carrying capacity or saturation limits are reached—a new paradigm we define and present examples of in this paper. [. . .] We suggest that if pulsing is general, then what is sustainable in ecosystems, is a repeating oscillation that is often poised on the edge of chaos. (547)

This influential paper sent legions of ecologists to look for the pulsing phenomenon elsewhere, and chaos ecology buoyed these efforts to understand the nuanced relationship between stability and disorder in nature over time. Carolyn Merchant's recent history of environmentalism includes chaotic ecology as one of only four fundamental approaches to the science (the others are human, organismic [population], and economic [systems]). In the twenty-first century, ecology gives the impression that balance in nature is really only a temporary state of poise between inevitable rapid changes in the system. Merchant writes, "The chaotic model of nature allows for the full expression of nature as an actor and shaper of history, rather than a passable backdrop to the inorganic machine. Unpredictable natural events and climatic conditions can trigger changes and transitions in local places, the impacts of which may be felt at great distances" (*American Environmental History* 116). Following the lead of the Odums, Merchant's humanistic perspective on the patterns of ecology imposes a narrative of complexity

that complicates classic environmentalism: "recent work in complexity theory characterizes a complex system as one that exists on the edge between
order and chaos. [. . .] Whereas an ethic based on the balance of nature
grants humans the capacity and power to restore degraded systems, chaos
and complexity theory challenge humanity to recognize nature as both predictable and unpredictable, orderly and disorderly" (190). Complexity, the
science of order and chaos in ecological systems, provides an intriguing new
avenue toward elucidating nature that goes well beyond classic reductionism. Although ecology built its foundation on the traditional scientific epistemology of simplifying complex systems down to their components, studying
parts of the system in isolation, and manipulating simplified models, the
new emphasis rests on narrating the flux among natural forces in their play
of perpetual dynamism.

This new paradigm has exciting potential to be developed through the
theories and methods of the humanities. After all, chaos ecology reanimates
classical mythology as a scientific narrative. Donald Worster explains the
powerful effect of character and narrative on the scientific imagination:

> If the ultimate test of any body of scientific knowledge was its ability
> to predict events, then the sciences, despite so many grand successes,
> were frequently failing the test. Making sense of that failure was the
> mission of an altogether new kind of inquiry calling itself the science
> of chaos. [. . .] Nature was far more complex than they had ever real-
> ized, or indeed, some were beginning to hint, than science ever could
> realize. Chaos was, like Gaia, a word that came welling up from the
> lost pagan cosmology of ancient Greece to seize the imagination of
> avant-garde scientists. If the earth goddess had long ago brought life
> and order into existence, then chaos had been her opposite: the realm
> where disorder still ruled, a dark underworld. [. . .] The scientific
> study of chaos began [. . .]. (406–7)

The imaginative power of the opposition between chaos and cosmos, two
faces of Gaia, enabled late twentieth-century ecologists to gain a conceptual
frame for the weird, vacillating data they were gathering. This neocatastro-
phism, updated from eighteenth-century geology, fulfilled cultural as well
as scientific requirements to explain the rapid changes in ecosystems. By
invoking historical literature, chaos ecologists were able to draw a portrait
of nature quite unlike the delicate perfect beauty that nature had been for

generations past. But characterizing nature has implications far beyond ecological science; it immediately becomes an ethical environmental practice. Frederick Buell discusses the ethical hazards of a chaotic worldview:

> A new philosophy of nature was used to justify a social system that not only intensified its environmental destructiveness but also erased many if not most of the more important rationales for guarding the health of the nature it was supposedly so intimately connected to. Chaos theory exposed the supposed fallacy of former conceptions of nature's health as equilibrial. It deprived environmentalists of appeals to the ideal of stable balance in nature and between nature and society. It also undercut the logic of nature protectionism by asserting that since nature was in continual disequilibrium and change, no form of existing nature ever was "pristine" or "primordial"—something that supposedly well-nigh invalidated all forms of wilderness protection. Most of all, chaos theory seemed to welcome, not be threatened by, environmental crisis; it made crisis itself seem creative. (207)

Although these points are well taken, Buell's analysis of the apocalyptic mode of storytelling as an environmental ethical disaster may be unduly pessimistic. Buell claims that chaos theory makes human impacts on ecosystems just another natural disturbance. He sees an irony in the idea that apocalyptic tales actually cause apathy and inaction because people told harrowing tales of the coming disaster surrender to the idea of mass extinction (201). On the contrary, perhaps narratives of chaos have greater power to increase popular awareness in today's anthropogenic environment because they are interesting to a general audience that enjoys seeing "good versus bad" characters compete in chaotic scenarios. Violated Nature and her human crusaders are justified in revenge. The egocentric villains whose sole object is to make a buck are common to this new popular discourse; they have been at least since the 1960s resurgence of American environmentalism and books like Edward Abbey's *The Monkey Wrench Gang* (1975). There is a clear distinction between natural and human disturbance, between the ecological idea that Glen Canyon experiences stochastic extreme droughts and floods rather than consistent annual flows, and the ethical idea of the Colorado River corked by a 220-meter-tall concrete wall so as to supply electricity to Sunbelt cities. Only extreme partisans confuse the issue. Narrative has the power to distinguish ecological theories from environmental ideas, but it can also reconcile scientific with humanistic methods in the understanding of our actions in nature.

Environmental historians tell stories. In their work, narrative plays a prominent role that upstages reductionism and systems modeling to explain ecological dynamics. These stories affect the public's vision of particular environments, and especially of how humans have used these places both at particular points in time and over wide swaths of human history. Narrative has greater respect for holism than do the mechanist or systems approaches, and it supports an imaginative approach to science. William Cronon is prominent among environmental historians who use a foundation of narrative theory to inform their work. Known for its carefully detailed philosophical histories of the midwestern United States during industrialization, Cronon's work sits at the crux between objective history and polemical treatise. Cronon is familiar with the postmodern critiques of false coherence and embedded value systems in foundational narratives, but he notes that narratives (unlike chronicles that impose no relationship between events) have the ethical power to make an audience care about the landscape at the center of concern. He argues for the motivational power in responsible narrative: "Despite the tensions that inevitably exist between nature and our narrative discourse, we cannot help but embrace storytelling if we hope to persuade readers of the importance of our subject. [. . .] Narrative is thus inescapably bound to the very names we give the world. Rather than evade it—which is in any event impossible—we must learn to use it consciously, responsibly, self-critically. To try to escape the value judgments that accompany storytelling is to miss the point of history itself, for the stories we tell, like the questions we ask, are all finally about value" ("Place for Stories" 1375–76). Though a skeptic might comment that "value" is most often imposed by the hegemonic culture that tells self-serving heroic stories, recent histories thoroughly critique the hegemony of industry and capital that has radically altered landscapes in the past two hundred years. Cronon notes that the same set of events can yield several stories that are utterly at odds when it comes to a sense of value, and his words echo theories of comedy and tragedy: "Stories are intrinsically teleological forms, in which an event is explained by the prior events or causes that lead up to it. [. . .] [I]f the tale is of progress, then the closing landscape is a garden; if the tale is of crisis and decline, the closing landscape (whether located in the past or future) is a wasteland. [. . .] A trackless waste must become a grassland civilization. Or: a fragile ecosystem must become a Dust Bowl. [. . .] However serious the epistemological problems it creates, this commitment to teleology and narrative gives environmental history—all history—its moral center" (1370). So closely tied to our cognitive process of learning, narrative has a unique facility in altering perspectives. Narratives that turn a critical eye on the values of the culture from which they have emerged, such as the narrative that renders the Dust Bowl a tragedy of human hubris, can cause serious problems for the status quo that directs disruptive operations. Morality is particularly powerful in environmental stories because it relates human activity to observed changes in nature. But it does not follow that nonteleo-logical narratives like Wells's *Time Machine* are inherently without a moral center. Narrating a story that is directionless and subject to chance events presents a cognitive challenge, but that challenge is the rallying center of postmodern narrative. Chaos ecology is also called postmodern ecology because of the energy and creativity drawn from these narratives of stochas-tic nature.

Postmodern ecologists recognize their science as a deeply historical prac-tice. For example, they question whether their field can accommodate the larger scientific pursuit of general laws and principles and apply them across the range of ecological conditions through time. Certain physical laws inhere in environments: water flows downhill, solar energy causes atmospheric mixing, condensation forms clouds. But more complex networks of causa-tion, such as population growth, community assembly, and species rich-ness, elude general principles and straightforward causal mechanisms because they are historical processes and may be approached at many levels of analysis (Cooper 218). Attempts to model these relations, such as the Lotka-Volterra equation of predator-prey dynamics, seemed for several decades of investigation to predict observed relationships, but Lotka-Volterra has more recently been exposed as flawed. There are too many divergent, unique, weird case studies in ecology for one equation to predict the outcome of them all. An enduring tension arises from the rivalry between theorists who seek general models for systems behavior and phi-losophers who use relative historical and narrative modes of representation (Kingsland 231). Models are epitomes of the general case, and postmodern ecologists argue not only that models are misleadingly mechanistic but that there is no way to derive general laws from radically situational and contin-gent entities like ecosystems. The only way to understand an ecosystem, a population, or a community is through a tailored story of origin, develop-ment, and chance.

The proleptic (future-looking) narrative of a changing environment has become a new genre in science writing that aspires to the realm of nonfiction. Stories about possible futures make climate change models readable. For example, "the model predicts 40 percent less rainfall in the wet season in the Sahel" turns by narrative into "the Sahara desert will continue to expand its geography southward and make agriculture and game hunting more arduous and less reliable in the poorest regions of Senegal, Mali, and Burkina Faso." Literary writers can take these narratives a step further and create fictional works about climate change; a novel might begin, "Binta Kinte had not seen an antelope in five years, though she lived somewhere within the old borders of Niokolo-Koba National Park. Her grandma's stories of the great herds that overran her village seemed as dreamlike as the legend that the Nieri Ko River hid man-eating crocodiles. Grandma's village was gone with all its antelopes, and the Nieri Ko was a dry snake that spat dust into the trickling Gambia." It is only two steps from computer modeling to forecasting fiction, and novels have greater power than quantitative models to foster public concern and care.

Chaos complicates narrative scenarios. Turning the predictive powers of models into narratives that envision the future is essential to seizing widespread public attention and inspiring changes in behavior and policy. A computer model might predict that sea levels will rise by one meter in the next century, but that prediction gains power only when a narrative draws out its consequences: coastal cities engulfed, arable land lost, violent hurricanes, the spread of disease, and so on. Alan Weisman's *The World Without Us* incorporates the predictions of many scientific disciplines in a coherent, wide-ranging story of what would happen on earth if humans disappeared. Weisman imagines the fate of different types of buildings, nuclear power plants, farmland, synthetic compounds, species biodiversity, and landmarks such as Mount Rushmore, which is carved out of stoic granite the features of which would remain recognizable for 7.2 million years (182). These narratives are more than scientifically detailed versions of the past as prologue, because they attempt to conceive of the unprecedented. Still, evidence of past upheavals remains our best basis for predicting the future in nonfiction narrative, and climate scientists look to the evidence of the major landscape shifts and mass extinctions that defined past geological epochs when they envision the next few centuries.

The novels explored in the previous chapters take a narrative approach to describing ecosystems in tumult, a perspective that allows them to juxtapose levels of impact across scale and time. Gilbert White's narrative achieves a decades-long perspective of environmental change through the accretion of quotidian impressions. Each anecdote he relates, such as the mother bird killed by the felling of her home oak tree, comes into relationship with higher-order events like deforestation in Selborne in the eighteenth century. White's innovative phenology, which looks for patterns of constancy such as the return of migratory birds at very specific intervals, is significant mostly in relation to violations of those patterns. The four narratives effectively involve the reader in radical environmental fluxes using the intermediary of a charismatic protagonist. These novels entertain wild manipulations of time (particularly in Shelley and Wells) and vivid accounts of the transformation of ecological place. Without narrative as the organizing, causational, and momentum-generating device, Gilbert White would only have compiled anecdotal lists of seasonal birds, Lionel Verney's epidemiological observations would lose their relationship to human extinction, Felix Aquila's postapocalypse fauna would be a merely fictional catalogue of taxonomic data, and the Time Traveler would have no distant evolutionary future to explore. The manipulation of conventional linear narrative into its fragmenting, pogo-stick condition in each of these works effectively conveys the many-leveled, interrelated, and yet chance-driven condition of the natural world, and particularly the ecologies of a human-impacted future.

The general public generally does not listen to ecological information until it appears in narrative form: Rachel Carson seized the public's attention with a narrative about the effects of industrial chemicals on watersheds in *Silent Spring*; James Lovelock observed global systems and constructed a sensational analogy starring Gaia; Alan Weisman wove together many threads of scientific observation to produce a bracing narrative of the posthuman globe in *The World Without Us*. By employing narrative, ecological science is able to move fully into public notice, where observations and directives are most powerful.

The U.S. Environmental Protection Agency frequently uses narrative descriptions of healthy and unhealthy ecosystems to complement numeric criteria that define a certain permissible concentration of a given contaminant. Written into the all-important Clean Water Act (CWA) is a series of "organoleptic effects" like taste and odor that can be used to describe the quality of a water body; unlike numeric criteria based on water testing, narrative criteria require the use of all the senses to detect the health of an ecosystem. These narrative criteria are not often set forth in elegant, artistic prose; in order to be as precise as possible, this discourse follows the legal model:

All waters, including those within mixing zones, shall be free from substances attributable to wastewater discharges or other pollutant sources that:

1. Settle to form objectional deposits;
2. Float as debris, scum, oil, or other matter forming nuisances;
3. Produce objectionable color, odor, taste, or turbidity;
4. Cause injury to, or are toxic to, or produce adverse physiological responses in humans, animals, or plants; or
5. Produce undesirable or nuisance aquatic life (54 F.R. 28627, July 6, 1989).

EPA considers that the narrative criteria apply to all designated uses at all flows and are necessary to meet the statutory requirements of section 303(c)(2)(A) of the CWA. (EPA)

Even where numeric criteria are established, policymakers who are often better trained in law than in ecology often ask scientists to state the narrative criteria implied by numeric thresholds, because narrative is so much more intelligible and compelling to a general audience. A narrative criterion is particularly useful when a given system is too dynamic or complex to be reduced to a set of numeric criteria that define, for example, maximum phosphorous and selenium concentrations. Still, in regulatory circles it can be hard to discern whether narrative is used to evaluate a system's health in the face of irreducible complexity, or whether, because of political and economic pressure, regulators are avoiding the establishment of "hard numbers" that might be enforced strictly, preferring a descriptive narrative interpreted differently by each reader. "Objectionable color, odor, taste, or turbidity" can describe an array of ecological conditions according to the loyalties of the interpreter, or "stakeholder," and subjectivity can justify inaction in a way that more objective numerical thresholds do not allow.

Despite its challenges, narrative continues to play a major role in environmental regulation. It is an essential descriptive tool that can be used to characterize complex ecosystems and protect them from abuse where numeric standards of contamination cannot be established. Narrative provides a holistic impression of ecosystem health that floats above a series of reductive models. Some postmodern ecologists claim that narrative is uniquely equipped to resolve inconsistencies in scientific modeling and provide intelligible protocols for further investigation and action:

> The power of science comes from the capacity of its narratives to convince us that something is general, and we should agree on it. And this agreement arises even when the story is quite long and encompasses inconsistencies. Furthermore, we seem to be able to agree even when there is no logical necessity in the outcome. We agree on evolution and global warming, even when many of the detailed models are at odds with each other [. . .] the story of anthropogenic global warming just feels right, and the science of it is confident. [. . .] The power of narratives, as with the power of myths, is their capacity to rise above contradiction, when the juxtaposition of large disparate issues is given meaning. (Zellmer, Allen, and Kesseboehmer 179–80)

An uber-narrative that describes a system in general terms and trends can unify superficial differences among conflicting models of systems. More dubiously, in the theory of Zellmer et al. it may not even matter whether the meaning created by narrative reflects external truth or merely constructs a façade of truthfulness; a simple consensus about meaning promotes changes in practice. Climate change skeptics seize on the different scenarios (stories) of future global trends as evidence that scientists are unsure about the phenomenon as a whole. This is a shadow of antievolutionists, who grasp at any disagreement among biologists about details of evolutionary mechanism as negations of the theory as a whole. Narratives can support general theories in science and gloss over differences in the details; models tend to foment rivalries among groups about those very details. We need narrative to hold together our common ecological stories about the past and future.

Literary scholars interested in ecocriticism should find themselves fully equipped to instill ecological principles, contradictions, and interpretations in a way that complements the work of their colleagues in the sciences. Ecology is not only a cluster of empirical strategies continuous with physics, chemistry, and biology; it is the most humanistic of the sciences because it is an interwoven fabric of landscape stories. Narrative is essential to the practice of prediction and is the voice of calls to action. It is capable of bearing the weight of predicting outcomes and instilling values to an inclusive audience. Mathematical chaos is a visual, rather than a narrative, practice that is demonstrated using fractals and weird, wonderful congruities in form. But ecological models of chaos rely on narratives of change in nature to be communicated. The chaotic story of anthro-nature meanders, branches, joins, and, we all hope, does not drop off the edge.

PART 2

Microcosm

4

Figures of Thought

Chaos performs the aesthetics of the sublime: the allure of the vast, mysterious, uncontrollable, and changing forms of nature. If chaos is sublime, the microcosm is a combination of the beautiful and the picturesque. It embodies the well formed, pleasing, and diminutive. It has an affective cause of calming human nerves, and a final cause of implying natural design in its exquisite organization. Microcosms are autonomous and complete. The eighteenth-century *concordia discors*—four vying elements in nature that paradoxically emerge as divine natural harmony—previews ecological microcosmology. The ancient Platonic concept of the human body as a microcosm evolved during the nineteenth century away from metaphysical ideals and toward material applications that eventually became formal experiments in ecology. One way the sciences draw patterns out of a chaotic world is to place boundaries around natural systems and study them in isolation from the larger world. Literature, I suggest, was an important mediator between the philosophical and ecological denotations of microcosm, because Romantic and Victorian poets developed microcosmic images as a method for exploring small-scale nature.

The microcosm is an epitome. It suggests that systems in nature can be modeled, and we can extrapolate onto the macrocosm simplified principles revealed by the model's behavior. The excess and variety of all cosmos become less bewildering when treated piecemeal as a series of semidiscrete systems. The microcosm has both mechanical and organic tendencies: it suggests that natural systems have an internal organization in which components play essential roles in the operations of the whole, but it also denies that these components can be extracted and fully understood as independent parts. The whole is a gestalt, and as an organic system, its complex functions dissolve under reductionism. Nature may be reduced to a set of subsystems that can be modeled, but ecology cannot atomize below this systems level without sacrificing the emergent interactive properties of the microcosm. In other words, the microcosm represents the lower limit of ecological reductionism. Critics of ecological modeling argue that microcosms have become too mechanistic and reduced to accurately represent organic natural systems, and too charismatic in their own right for the good of those natural systems: microcosms are lab pets that keep ecologists indoors. With these criticisms in mind, perhaps the literary microcosms discussed in the coming chapters will demonstrate the greatest synthetic potential between science and philosophy for this enduring figure of thought.

The microcosm is often more of an idealization than a clear-eyed representation. Desirable miniature worlds may be cultivated, as gardens are models of fecund nature, or engineered, as biosphere projects are constructed utopias. Romantic-era writers are nearly always beguiled by the ideal potentials of the trope, and early microcosm ecologists like Stephen Forbes follow suit with their framed, picturesque studies. Later in the nineteenth century, however, both literature and science would discover how the model is useful not because it is whole or perfect or pleasant but because it is most instructive about the state of larger-scale nature when endangered and degraded. In this chapter, I survey the earlier idealisms of microcosmology and explore how Romantic philosophy employed the imagination to demarcate and manipulate closed spaces within landscapes. The following chapter is devoted to the earliest hints of the microcosm in science, and the evolution of literary models toward a rhetoric of disturbance in the Victorian era. The third chapter in this section discusses the essential role of microcosms in ecology and the problems associated with reductive modeling in a holistic science.

The microcosm originated as a philosophical trope in ancient Greece. As an aesthetic device for organizing spatial relationships, cosmos was seen as the opposite of chaos, a figure of order and harmony and the sum total of human experience. The word "cosmetic" derives from the desire to beautify and harmonize disordered elements of human faces. To Hippocratic medicine, the microcosm was a figure for the human body, which displayed in miniature the elements and energies of an ordered cosmos. Illness, then, could be understood as an imbalance in this corporeal system, and philosophers before Socrates used the microcosm conceit to understand stability and variation both within the body and in relation to larger spheres, including immediate environments, whole bioregions, and the macrocosmic Earth. The four elements of earth, air, fire, and water corresponded with the four humors, and these analogical elements required a proper balance for the enjoyment of health in both human bodies and their corresponding physical environments.

The monarchal microcosm so well known to Renaissance writers is a political version of this medical conceit: the king's body represents his dominion, and the health of both spheres depends on the wise and self-disciplined purview of their leader. Shakespeare's *Richard II*, for example, develops the monarchal microcosm as a means of exploring how failures of authority can degrade the ecological state in parallel with the decline of the body politic. Richard's gardeners use the microcosm of the walled garden as a conceit for the energies required to maintain political control against parasitic rivals (3.4). Richard frequently laments the wasting of his physical body, but he is blind to the vehicle of the gardeners' microcosm metaphor, the degraded garden (Scott 267–71). In their Platonic understanding of interscalar relations, the gardeners identify Richard's nefarious advisors as fast-growing weeds that choke out his long-established monarchal tree. Renaissance metaphysics and Enlightenment science first became intrigued by the minute worlds just beneath our perceptions when scientific technologies like microscopy stirred extensive philosophical debate about interscalar correspondence in the cosmos. The "minute particular" is a fact revealed in the microscopic world that reveals structure at larger scales—part of Robert Hooke's argument that the microscope paradoxically caused the enlargement of perception (Chico 144).

Romanticism developed a philosophy that the imaginative brain was its own little world that could rework nature into distinctive harmonious forms. As a redress of the prevailing empirical science of the time, Wordsworth and Coleridge experimented with a unique natural perception by allowing the mind to direct and diversify sensual input, rather than having the senses impose on the mind a singular view of the world. This epistemological proposal influenced Wordsworth's "Home at Grasmere," and creates a psychological microcosm that revels in subjective immersion in nature. Though it may seem inimical to any scientific method of objective observation, Romantic philosophy carries a surprising affinity with future ecological methods. Both ecological science and literature treat microcosms as simultaneously theoretical and literal. Both value the aesthetic pleasure and epistemological insight that results from circumscribing heterogeneous, synergetic systems and controlling their fate. The brain-as-world-maker notion of Romantic philosophy proposed that the alchemical action of poetry might reverse the traditional "senses inform mind" into "mind directs senses." In order to preserve the memory of a place in nature, poets studied landscapes with such devotion that their brains half-created the nature they knew so well, as when Wordsworth describes the mind as a "mansion for all forms," or when Keats, in his "Ode to Psyche," creates a garden in "some untrodden region" of his mind and decorates his garden with "the wreath'd trellis of a working brain" (51, 60). Coleridge was familiar with the microcosm in philosophy through his reading of the German Romantics Schelling and Novalis, who claimed that the microcosmic man could envision and revive a better relationship between nature and humanity (Halmi 49). By imaginative manipulation, Coleridge transforms his lime-tree bower from a prison into a garden. Coleridge defines the poetic symbol as that which "partakes of the reality which it renders intelligible; and while it enunciates the whole, abides itself as a living part in that unity of which it is the representative" ("Statesman's Manual" 619). The symbol is an epitome within a larger complex living system, a model that gives a glimpse of the ineffable whole.

Cognitive science applied to literature has recently shed light on the relationship between contemporary understandings of the brain and poetic production, but no literary critic has yet identified the Romantic conceit of the mind as a microcosm. This psychological microcosm provides a new view of Romantic aesthetics that relies not simply on the sublime massiveness of the macroscopic in nature, or on the beautiful minuteness of the microscopic. Instead, the trope constructs scalar parallels between various structures in nature, and the brain of the perceiver serves as catalyst for epiphanies of ecopoesis. Ecology and poetry share the conviction that a small system, if designed properly, can model many potentialities of ecological organization. Where writers control their systems through image and prosody, scientists control microcosms by manipulating physical parameters through many permutations of experiment.

The present chapter is devoted to showing how Romantic literature imagined the ecological microcosm before science adopted the microcosmic sphere as an experimental strategy. By looking at the microcosm through a literature of environmental engagement such as the nineteenth century offers, we come to appreciate how advances in concept are often borne on free imagination before they can be formalized and fully understood as effective models for generating scientific data. All of the poems discussed in this chapter use the microcosm as an orientation point between the writer and the natural world. Romantic-era writers figured the microcosm ecologically, that is, as a cognitive lens through which to envision nature. Some Victorian-era writers, especially women, became skeptical of the egotistical, brain-fetish mode of the male Romantics. The satirical microcosms of George Eliot and May Kendall critique the mere imagining of natural utopias while real landscapes fall under industrial machines, as we will see in the next chapter. Because nineteenth-century poetry so frequently turns to despoiled nature under the forces of industrial change, the canon portrays various nascent perspectives on how the trope can be used ethically. Microcosms are often simultaneously ecological and environmental.

While my argument makes no causal claims that would artificially promote an essential mediatory role for literature between Platonic philosophy and ecological science, the evident fascination with minute natural systems in nineteenth-century literature represents British culture's move toward the holistic empirical thinking that would be necessary for ecological science to fledge by the turn of the twentieth century. For it to be useful in ecological science, the microcosm had to evolve away from its origin as a philosophical theory toward a controlled empirical scheme set within a material system. Between these two distinct but related ideas lurks an intriguing study of epistemological evolution. The scientist Stephen Forbes, admitting that he had not the poet's talent for *ekphrasis*, still adopted a poet's perspective to introduce the aesthetic of the lake as a microcosm to a scientific audience for the first time in 1887. This chapter proposes several literary origins for his act of consilience.

Romantic Models of Ecological Systems

What follows is a series of interpretive vignettes on small natural spaces in Romantic poetry. Nearly a century before Arthur Tansley theorized the ecosystem for the sciences, poets were noticing how their natural dwellings could be imagined as living subjects. Though neither exactly prosopopoeia, speaking for the landscape, or theriomorphism, figuring a human as an animal, poets began to encircle spaces of nature as quasi-organisms. With a microcosm as a poetic subject, nature became embodied in a model; poets studied these lyric spaces as ecologists would later study ecosystems. With a desire to reinfuse imagination and spirit into Enlightenment nature, and under increasing industrial stresses on the English landscape, Romantic writers actuated the power of the ecological microcosm as a material construct built upon an imaginative conceit. These literary encounters actively recombined some parcel of physical nature with the writer's imagination of her place within the cosmic net. These poems are half within nature itself and half within the mind of the perceiver. These literary ancestors to modern ecological science are, indeed, works of art. Literature since at least the metaphysical poets has negotiated with scientific theories about how nature behaves. In earlier chapters, we observed how narrative dynamics can evoke rapid environmental change, a signal pattern of contingency in postmodern chaos ecology. The present chapter shows the symbolic sense of control and pleasure afforded by a simple system that models the global macrocosm, and also the vulnerability to utter dissolution that models can exhibit, portending ecosystem upheaval.

The Romantic aesthetic of the picturesque used the frame to create a subset of nature. The vogue of framing picturesque landscapes using the Claude glass demonstrated the popular desire to recuperate the Edenic bower by isolating a scene, and thereby symbolically possessing nature while at leisure as a tourist. In 1794, William Gilpin's essay on picturesque beauty formalized the notion of composition as "uniting in one whole a variety of parts; and these parts can only be obtained from rough objects [. . .] the picturesque eye is to *survey nature*; not to *anatomize matter*. [. . .] It examines *parts*, but never *particles*" (508; original italics). Gilpin contrasts the reductionism of the anatomizing physical sciences, yet still seeks a limiting frame for his formal vision of the picturesque. He requires the partition of nature for aesthetic effect, noting his opinion that "her ideas are too vast for picturesque use, without the restraint of rules. [. . .] [The painter must] remove little objects, which in nature push themselves too much in sight, and serve only to introduce too many parts into your *composition*" (510). Simplicity, variety, and synergy became the requirements for viewing a landscape properly and for sampling a piece of nature as an idealized representation of the whole. Coveted artistic taste relied on isolating the essential aesthetic elements from the hyperabundance of nature as a whole.

Writers of Gilpin's era frequently mocked the vogue of the picturesque, but not as often as they used it. Elizabeth Bennet flippantly dismisses the Bingley sisters and Darcy because her presence on a walk at Netherfield would "ruin the picturesque" of three in a hedgerow frame, but Austen employs Gilpin's language without irony to describe Pemberley through Elizabeth's eyes. Coleridge lamented the "pikteresk toor" that brought legions of carriage-borne invaders to the Lake District, but his companion Wordsworth published an enabling and popular *Guide to the Lakes* that, along with the railroad age, greatly escalated local tourism. For the Romantic poets, the microcosm is the world of the mind imaginatively reflecting nature. Wordsworth's epiphanies, his "spots of time," appear when emotion is recollected in tranquility, when the blizzard of experience is distilled by the brain down to a few symbolic images. The preamble to the "Intimations of Immortality" ode describes its philosophical foundation. Wordsworth as a boy was overwhelmed by the solipsistic "abyss of idealism" that his imagination continually created, and he sought escape by grasping at "a wall or a tree," something material in nature that was definitively outside his mind (*Complete Poetical Works* 5:52). This memory leads to the poem's philosophic occasion: "Archimedes said that he could move the world if he had a point whereon to rest his machine. Who has not felt the same aspirations as regards the world of his own mind?" (5:53). Poetry is the vehicle that moves the whole cosmos of the psyche. Unlike religion, which appeals to spirit and soul, or science, which appeals to reason, poetry for the Romantics was a wellspring for moving the mind all at once. Poetry did not have to reduce the whole into a ration of spirit, imagination, intuition, emotion, memory, and reason; it was the cistern containing an "unentangled intermixture," as Shelley wrote. Over the course of his epic poem *The Prelude*, which was to rival the world-wandering heroic epics of past ages, Wordsworth defined his scope in a microcosm with the subtitle *or, Growth of a Poet's Mind*. Wordsworth's brain distills the natural world into "spots of time," episodes encapsulated by the rhythms of blank verse.

An iconic image of literary Romanticism is preserved in Caspar David Friedrich's painting *The Wanderer Above the Mists* (1818), which shows a hiker on an Alpine pinnacle gazing down on a sea of fog pierced by jutting crags. This scene is echoed in the culmination of Wordsworth's *Prelude* as he climbs Mount Snowdon, a poem that I cite throughout in the more liberated 1805 edition: "I found myself of a huge sea of mist, / Which, meek and silent, rested at my feet: / A hundred hills their dusky backs upheaved / All over this still Ocean, and beyond" (13.43–46). As he meditates upon the day's visions, Wordsworth recalls the iconic mountain scene as "The perfect image of a mighty Mind, / Of one that feeds upon infinity" (13.69–70). The imagining "Mind" inspired by the eternal Alps, feeding upon both apparent forms and the mysteries of the obscuring fog, finds its counterpart in the bold and "lonely Mountain" (13.68). In these spots of time, Wordsworth sees poets as the "Prophets of Nature" who "speak / A lasting inspiration, sanctified / By reason and by truth" (13.442–44). Arguing against the common Enlightenment critique that poets create flighty works out of pure fancy, Wordsworth fuses inspired imagination with the sublunary offices of reason and truth, which corrals the best of all knowledge in the poets' corner.

He is then qualified to "Instruct [men] how the mind of man becomes / A thousand times more beautiful than the earth / On which he dwells, above this Frame of things / [. . .] In beauty exalted, as it is itself / Of substance and of fabric more divine" (13.446–48, 451–52). The static face of nature is animated by the mental activity that stirs up life in dead forms and absolves aesthetic strife. Crucially, the exercise makes *the mind* more valuable, more beautiful, not the forms of nature that catalyze the inspiration. This highly anthropocentric resolution follows from the first book of *The Prelude*:

> The mind of man is fram'd even like the breath
> And harmony of music. There is a dark
> Invisible workmanship that reconciles
> Discordant elements, and makes them move
> In one society. [. . .] (I.351–55)

Dividing mind from nature causes personal discord, as Wordsworth expresses in his grim passages on the spiritual pollution of cities. Harmony results when the brain works the natural elements into synergy. This cognitive alchemy allows the poem to develop its natural spiritualism simultaneously in the spheres of the poet's mind and in the adventures in nature that are his ministrations. The "Invisible workmanship" of the mind comes in as Nature "would frame / a favor'd Being" by imposing both "gentlest visitation[s]" and "Severer interventions" (1.367, 370). If nature is a chaos of clashing elements, Wordsworth proposes (in company with Kant) that the brain possesses organizing principles that build purposiveness and teleology into the narrative. If the discord is cognitive, nature promotes harmony by providing memorable experiences that live on in the poet's mind and become symbols of larger meaning. Synergy emerges through an inward-outward exchange in which nature both nurtures and tests the perceiving mind, and the mind concocts a more beautiful, organized natural world than actually exists in the mere matter of the elements. This subjective view of true perception is typical of Romantic poetry, and it carries surprisingly into the early formulations of the microcosm in ecology, where, for example, Stephen Forbes's microcosm (1887) is found only "within the mental grasp," and Arthur Tansley's ecosystem (1935) is a "mental isolate" rather than a purely physical structure that could be perceived objectively.

In her study of literary Darwinism, Gillian Beer observes the use of "lyrical materialism" by the elder Romantics, which leads to a secular-spiritual way of knowing nature: "Wordsworth's emphasis upon the congruity of the inner and the outer worlds allows harmony and development without the need to insist upon a preordained design" (45). In an echoing passage from "Tintern Abbey," William recognizes in nature a spiritual energy that is "anchor of my purest thoughts" (line 110), and wishes for Dorothy to grow, intellectually, to resemble himself: "thy mind / Shall be a mansion for all lovely forms, / Thy memory be as a dwelling-place / For all sweet sounds and harmonies" (140–43). Dorothy's mind and memory did indeed form around her natural experiences with William and others, and a look at how microcosm figures in her poetry reveals that she was often more scientific and exact in her observations of nature than her brother. William's insight is that worlds can be engulfed and arranged in the architectural mansion-mind. The poet's brain is a ravenous organ that contains, arranges, and harmonizes nature. Keats inherited this eco-psyche notion with his idea of human experience as a "mansion of many apartments" that light up with innocent pleasure and darken as the individual gains experience of "misery and heartbreak" and feels acutely "the burden of the mystery" (Selected Letters 124). Keats quotes Wordsworth but subverts his enlightened moment in "Tintern Abbey," where Wordsworth's "burden of the mystery" is "lightened" by transformative contemplation of nature. For Keats, experience is darkening while it is enlightening. All three writers, and many of their contemporaries, are interested in the container of the brain and how nature can be manipulated in cognitive space.

Wordsworth's "Home at Grasmere" is a rewarding study in the context of an ecological microcosm as a dwelling place. The long poem was meant as a first section to his unfinished epic "The Recluse," but the introductory fragment was not published until 1888, long after his death. Wordsworth felt that he had failed on the "Recluse" project, but "Home at Grasmere" and "The Excursion" (the epic's second section) together form an appealing rejoinder to his finished epic *The Prelude*. In "Home at Grasmere," Wordsworth set out to establish a philosophy of organicism between the mind and its natural dwelling. Coleridge recalled that "Wordsworth should assume the station of a man in mental repose, one whose principals were made up, and so prepared to deliver upon authority a system of philosophy. He was to treat man as man,—a subject of eye, ear, touch, and taste, in contact with external nature, and informing the senses from the mind, and not compounding a mind out of the senses" (*Table Talk* 36). Wordsworth's verse précis of this philosophical occasion is more inclusive of a two-way exchange of influence:

> my voice proclaims
> How exquisitely the individual Mind
> (And the progressive powers perhaps no less
> Of the whole species) to the external World
> Is fitted; and how exquisitely, too—
> Theme this but little heard of among Men—
> The external World is fitted to the Mind;
> And the creation (by no lower name
> Can it be called) which they with blended might
> Accomplish; this is our high argument.
> ("The Recluse" 815–24)

As philosophers who so deeply valued the imagination's contribution to our perception of the material world, the pair of poets set out to reverse the conventional order of a plastic mind molded by its concrete habitat, substituting (without fully eliminating the obverse) an ecological habitat created by an active brain. If nature is plastic and organic, not concrete and mechanical, it is aesthetically in harmony with the creative imagination: it can be seen as an evolving *natura naturans*. For Wordsworth, more can be known of nature by recombining it with imagination than could be known by classical scientific objectivity. This characteristic both enables and stifles

Wordsworth's ecological usefulness; as James McKusick has noted, "his poetry is best regarded as an evocation of lived experience, rather than a scientific description of the natural world" (25). The Romantics' assertion of the truth of imagination as its own legitimate form of knowledge allows their poetry to enlarge the epistemology of the sciences. The sphere of possible knowledge expands from observation and cataloguing to the more personal but universal modes of memory, emotion, desire, and projection. As Milton's high argument had been to "justify the ways of God to men," Wordsworth performs the more modern task of justifying the ways of nature to men by elevating human agency through imagination. Nature is half-forged in the folds of the thinking brain. It is a naturalist's version of Hamlet's "there is nothing either good or bad but thinking makes it so" (2.2.268–70).

Wordsworth's habitat is the microcosm of Grasmere vale. Though Wordsworth is what Schiller called a sentimental poet, a nonnative who is aware of the rift between real and ideal, there is very little irony in the idealized sense of dwelling and belonging in Grasmere's bounds. The vale is companion and counterpart to the heavens reflected in its waters, and Grasmere isolates and distills the perplexing excesses of all nature down to a clarifying unity in which the mind could dwell and make daily exchange:

> feeling as we do
> How goodly, how exceeding fair, how pure,
> From all reproach is yon ethereal vault,
> And this deep Vale, its earthly counterpart,
> By which and under which we are enclosed
> To breathe in peace; we shall moreover find
> [.................................]
> The Inmates not unworthy of their home,
> The Dwellers of their Dwelling.
> (639–44, 647–48)

A dwelling must have dwellers truly to exist; otherwise, the dwelling is merely a symbol of what is lost, a notion Wordsworth explored in "The Ruined Cottage." Grasmere is a dwelling both physically and cognitively, and is vital by virtue of the inmates (human and otherwise) who imaginatively perceive its exquisite organization. The work of achieving congruence between mind and nature, in effect, settles and coheres both entities; the thesis and antithesis are mutually enabling toward a desired synthesis in which nature and the mind are mimeses of each other, and neither holds precedence. The mind isolate is a vale of nightmares:

> Not Chaos, not
> The darkest pit of lowest Erebus,
> Nor aught of blinder vacancy, scooped out
> By help of dreams—can breed such fear and awe
> As fall upon us often when we look
> Into our Minds, into the Mind of Man—
> My haunt, and the main region of my Song
> ("The Recluse" 788–94)

However, nature chastens the mind into a harmonious union:

> [. . .] the discerning intellect of Man,
> When wedded to this goodly universe
> In love and holy passion, shall find these [Elysian fields]
> A simple produce of the common day.
> (805–8)

"Simple" and "common" products of nature reinfuse quotidian experience with the revelations of the poetic mind wedded to its natural dwelling place. The human mind in a state of ennui is more awful than mythical chaos, but it finds sublime transcendence when united by ecological nuptials in "Home at Grasmere." Wordsworth is self-consciously performing his own intellectual wedding: "Embrace me then, ye Hills, and close me in; / Now in the clear and open day I feel / Your guardianship; I take it to my heart; / 'Tis like the solemn shelter of the night" (110–13). Now husband to his circumscribed piece of nature called Grasmere vale, Wordsworth's mind and Grasmere make each other nobler by deep appreciation, and the microcosm gains the emergent personable qualities of nurturer, companion, instructor, and guardian.

The ecosystem was to become science's conceptual theater in which the drama of life plays out, the space in which individuals within communities grow, strive, learn, find provision and in turn provide. Wordsworth anticipates the scientific drama of evolutionary ecology with his own lyrical drama of cognitive evolution set within a natural compass. Though the voice is necessarily self-centered, he frequently addresses the ecstasy of other figures, the flora and fauna that reflect his personal joy with their own effusions. The holistic default of "Perfect contentment, Unity entire" fuses the disparate parts of Grasmere into a single ecosystem (line 151). He arrives at this phrase very early in the poem. It is an ideal that he will seek for the whole work as a mantra reinforced by its very repetition. The phrase is a keystone to a long philosophical perambulation:

> What want we? have we not perpetual streams,
> Warm woods, and sunny hills, and fresh green fields,
> And mountains not less green, and flocks and herds,
> And thickets full of songsters, and the voice
> Of lordly birds, an unexpected sound
> Heard now and then from morn to latest eve
> Admonishing the man who walks below
> Of solitude and silence in the sky?
> These have we, and a thousand nooks of earth
> Have also these; but nowhere else is found
> Nowhere (or is it fancy?) can be found
> The one sensation that is here; 'tis here,
> Here as it found its way into my heart
> In childhood, here as it abides by day,
> By night, here only; or in chosen minds
> That take it with them hence, where'er they go.
> —'Tis, but I cannot name it, 'tis the sense
> Of majesty, and beauty, and repose,
> A blended holiness of earth and sky,
> Something that makes this individual spot,
> This small abiding-place of many men,
> A termination, and a last retreat,
> A centre, come from wheresoe'er you will,
> A whole without dependence or defect,
> Made for itself, and happy in itself,
> Perfect contentment, Unity entire.
> (126–51)

Wordsworthian aporia pushes this passage carefully to its conclusion. The inquisition of his own happiness ("have we not?"; "is it fancy?"; "I cannot name it") creates pauses in which his perception catches up with his intuition before the wedding can move on (26, 36, 42). Though "a thousand nooks of earth" are other microcosms possessing equivalent natural virtues, Grasmere is unique to Wordsworth because it holds the center and circumference of his cognitive, emotional, and physical being (34). His chosen vale outshines all others not because it is objectively exceptional but because it is the locus of his emotive attention; it is his little world. Reality itself is constituted by the union of ideas and evidence. In science, such weddings (when well corroborated) grow to the status of theory: a collection of concepts, both abstract and observable, uniting to form a principle that best explains a class of phenomena. Wordsworth's ecstatic phrase "Perfect contentment, Unity entire" resounds off the encircling hills. This nook of nature is coherent, self-sufficient, reducible without splintering into atomies, capable of providing both contentment (emotional) and unity (structural). Wordsworth's microcosm is a utopian vision, embellished by a mind set in reverie and drawn along by iambs, but it is more than mere lyrical idealism. It sets a circumference around the mind-in-nature, and foregrounds the formalization of insular, unified systems in empirical studies of ecology.

Dorothy Wordsworth has historically been known as a companion and chronicler for her more famous brother whose observations on nature would emerge, reformulated and lyricized, in William's poetry. Dorothy's notable ecological visions are often based on a desire for insulation within the pleasurable vales of a cultivated landscape. This desire lies at the heart of bioregionalism, the valuation of self-sustaining and autonomous ecosystems driven by a deep investment in place.[1] However, Dorothy is more successful than her brother as a close observer of material nature, with less embellishment and more particulars, as her notebooks from Grasmere prove. Her eye for empirical detail renders two of her poems poignant reflections on small-scale nature, and both are nostalgic *ubi sunt* perspectives on lost worlds. The first, "Irregular Verses," addresses the daughter of Dorothy's close childhood friend Jane Pollard. Dorothy describes the life she had imagined living with Pollard before the women were "by duty led" down separate paths. Their lost scheme involves a self-enclosed, self-sustaining feminine space of nature:

1. Cervelli's Dorothy Wordsworth's Ecology is a major recent study of the Wordsworths' strategies for depicting the natural world, though microcosms and circumscription are not part of Cervelli's argument.

A cottage in a verdant dell,
A foaming stream, a crystal Well,
A garden stored with fruits and flowers
And sunny seats and shady bowers,
A file of hives for humming bees
Under a row of stately trees
And, sheltering all this faery ground,
A belt of hills must wrap it round
[. . . . .]
Such was the spot I fondly framed
When life was new, and hope untamed.
(21–28, 35–36, quoted in Levin 201)

This Georgian idyll is informed by a picturesque aesthetic, with the observer having "framed" a subset of nature as the salubrious balance of sustenance and beauty. Significantly, this habitat is an imagined world, not one that Dorothy ever inhabited. Though only a figure of fancy, this "faery ground" is an archetypal description of an ecological microcosm. The domicile of the cottage broadens its circle to include the immediate natural surroundings, and her study of a dwelling place grows to encompass a bioregion.

Dorothy's utopian vision transcends fantasy because it is ecologically precise. She details the biodiversity necessary to natural systems by proceeding along a series of enabling oppositions: cottage and dell, stream and well, sun and shade, garden and trees. The synergetic system is closed by a "belt of hills" that hold a feminine grace implicitly opposed to masculine mountains like William's Alps. William's "Home at Grasmere" celebrates the dwelling of the mighty mind, "the *sense* / Of majesty, and beauty, and repose, / A blended holiness of earth and sky" (142–44), but it mumbles over the details, the "flocks and herds, / And thickets full of songsters, and the voice / Of lordly birds" (128–30). With William, thickets and songsters may remain unspecified. James McKusick has noted that William "tends to subordinate the description of nature to the inward explorations of poetic self-consciousness. [. . .] [T]his imaginative power is fundamentally at odds with the detailed perception of flora and fauna, or the discovery of their complex interrelations" (25). The locus of his attention is the mental alchemy of the whole assembly of living landscape. His sister's "Irregular Verses," by contrast, celebrates the picturesque composition as a careful articulation of natural dynamics. Each species has a home: garden, cottage, and hive; each species has a role: to photosynthesize, to cultivate, and to pollinate. Where William uses the landscape to demonstrate the open exchange between natural grandeur and the poet's evolving mind (prompting Keats's critique of the "egotistical sublime"), Dorothy's vision is purely occupied with the ecosystem. There would be no home without fresh water, food, and shelter. There would be no fruit without flowers and bees. The seats and bowers for human figures are empty, but everything else works together to maintain the classical balance of nature. William is the center of his scene, and his literal self-centeredness is an attractor that affects the natural system dynamics. Dorothy is absent from her imagined scene, and her absence permits a more devoted perception of its independent structures.

A second poem of Dorothy's speaks to the ecological microcosm. "Floating Island at Hawkshead, an Incident in the Schemes of Nature" was published in a volume of William's poetry in 1842. This island floated in the subconscious of the sibling poets for decades: William attended Hawkshead in the 1780s, and Dorothy's poem is thought to have been written in the late 1820s, though it was not published for another twenty-two years. William likens himself to the drifting island in his *Prelude* of 1805 (13.339–43). He follows in the Shakespearian tradition of the "wandering bark" as a metaphor for the seeker looking to the stars for direction. Dorothy works the floating island quite differently: she allows it to remain its own literal entity, as a proof of how "Harmonious Powers with Nature work" ("Floating Island" 1). Rather than impose the poet-seeker on the island's condition, she describes the coherent vitality of an autonomous system. The island is

> Dissevered float upon the Lake,
> Float, with its crest of trees adorned
> On which the warbling birds their pastime take.
>
> Food, shelter, safety there they find:
> There berries ripen, flowerets bloom;
> There insects live their lives—and die:
> A peopled *world* it is;—in size a tiny room.
> (10–16)

From her perspective, the island's very existence demonstrates the small portions in which nature can remain whole, interactive, and interdependent. By retaining the heterogeneity intrinsic to a healthy system, the island also models the whole. It is "peopled" in the older sense of the word: inhabited by living creatures that need not be human to lend a sense of biological civilization. This is not a barren place; it is a "peopled world" that is both literal in itself and evocative of larger nature. Insects' lifetimes are contained in this tiny space. Birds linger, eat berries, and presumably fertilize the soil and infuse it with seeds. In both poems, Dorothy lashes together homologues: Trees-flowers-berries; birds-insects-people; food-shelter-safety. Flora, fauna, survival.

In the end, the floating island is a mortal superorganism, both self-sufficient and vulnerable. The poem ends with an assertion of the island's materiality: "Its place [is] no longer to be found, / Yet the lost fragments shall remain, / To fertilize some other ground" (26–28). Its substance is recycled into the bottom of the lake, its genetics distributed by birds on the wing. It is at once an empirical object evincing ecological structure and a figurative study of worlds within worlds. The poem successfully joins poetic imagination with nature's material systems by virtue of its small scale.

Romantic poetry is equipped with powerful lenses of the imagination. These lenses navigate the scales of nature by panning in to particulars and out to universals. Traversing scales in nature, in effect, prepares the poetic mind for its moment of epiphany. Cognitive chemistry is involved in transforming the mundane into the transcendent, the quotidian into the symbolic. In "Kubla Khan," Coleridge's laudanum-driven verse describes the ideal, circumscribed space of paradise:

> So twice five miles of fertile ground
> With walls and towers were girdled round;
> And there were gardens bright with sinuous rills,
> Where blossomed many an incense-bearing tree;
> And here were forests ancient as the hills,
> Enfolding sunny spots of greenery.
> (6–11)

This walled garden paradise is an exotic Eastern Eden. Here, Coleridge's "psychological curiosity" refigures Eden using the dreaming imagination and the opiate as vehicles. Coleridge claimed he lost his memory of the rest of the poem because he was interrupted in his frenzied composition by a knock at the door. This Xanadu existed tenuously in his mind only while his neurons held the short-term memory of the vision that vanished, as

Coleridge says in the poem's preamble, like "images on the surface of a stream into which a stone had been cast" (*Complete Poetical Works* 212). Like half-remembered dreams, microcosms of the mind are vulnerable to dissolution if external matters intrude. Still, Xanadu encompasses a complete, complex ecosystem: river, sea, fertile ground, gardens, forests, sunny spots of greenery, romantic chasm, green hill, cedarn cover, fountain, rocks, wood and dale—all contained within a sunny pleasure-dome.

In a fifty-four-line poem packed with tense and erotic drama, it is easy to overlook the remarkable articulation of ecological complexity and the landscape's division between synergetic balance and chaotic change. The irregular dense rhyming and the driving rhythm of the poem make the lines an interdependent system that holds itself together, but only barely. Iambic tetrameter yields to pentameter, which is interrupted by the chorus of a damsel with a dulcimer. Quatrains and couplets bristle with slant rhymes. The chasm fountain splits the balanced, stable five miles of fertile ground with huge fragments of airborne rock. Pastoral meets apocalyptic; rhapsody meets jeremiad. Xanadu is a symbol of ecological extremes, utopia and dystopia wrapped up in a bundle of energetic verse. The Georgian eclogue of "twice five miles of fertile ground" ruptures under the geological force of the "earth in fast thick pants" (18)—a striking preview of plate tectonics and geothermal activity, where the pressures of subterranean "ceaseless turmoil" force explosions of geysers and rock (17). Space is fraught with contrast between the picturesque "gardens bright with sinuous rills" (8) and the sublime "caverns measureless to man" (4). The spontaneous river flows "Five miles meandering with a mazy motion"—an image that implies a lateral five miles indefinitely extended by the crazy path of the water (25).

The poem is so tightly wrapped in its own bizarre perversity that even realistic details feel fantastic. This small world of its own making, pushed to the limits of its garden walls and cranial cavern, models diverse ecological conditions along several axes: sunny and sunless (energy), gardens and forests (land use), and hills and caverns (topography). Coleridge's caverns are in part the bottomless depths of the opiated psyche, but they are also all unknowns in science that remain "measureless to man." The "lifeless ocean" that receives the "sacred river" may only be lifeless because it is yet unknown; cartographers have often conflated terra incognita with barren or desert land (28, 26). Trying to recall the divine music of the dulcimer, the adventurer declares, "I would build that dome in air, / That sunny dome! those caves of ice!" (46–47). The dome contains the microcosm analogously as the sky contains the earth, or, as Coleridge writes in "Christabel," "the blue sky bends o'er all" (line 319). Along with the dream vision, the model of Xanadu breaks apart in wakefulness and leaves a residue of forbidden Paradise on the reader's mind. Coleridge's chaotic microcosm is a vision of a brain-world in tumult, weird and wild, completely untethered from the domestic niceties of the Wordsworths' small worlds.

Coleridge's poem "This Lime-Tree Bower My Prison" is a more sober study of natural scale and the power of alchemical imagination. The poem wrestles with the mundane until it finds its moment of insight, but the poet has no need to climb a mountain peak or take laudanum for his perspective. Simply by dwelling in the same small, complex, natural place and cogitating through his mood, Coleridge transforms blindness and separation into a vision of interconnection. An invalid with a bum leg, Coleridge sits beneath the lime tree and imagines his friends' progress on their walk along the hilltop, down the dell, past the waterfall, and back into the upland meadows. He is moved to empathy and transposes their varied experience into his own circumscribed location:

> A delight
> Comes sudden on my heart, and I am glad
> As I myself were there! Nor in this bower,
> This little lime-tree bower, have I not mark'd
> Much that has sooth'd me. Pale beneath the blaze
> Hung the transparent foliage; and I watch'd
> Some broad and sunny leaf, and lov'd to see
> The shadow of the leaf and stem above
> Dappling its sunshine!
> (45–53)

Coleridge finds in the sun patterns on a single leaf a recursive view of an entire grove of sunbathed trees. He harnesses his imagination into empathy with Charles Lamb, whom he envisions gazing "on the wide landscape" (40), and his active brain transforms his own small scene into a vision of nature's scalar congruence:

> Henceforth I shall know
> That Nature ne'er deserts the wise and pure;
> No plot so narrow, be but Nature there,

No waste so vacant, but may well employ
Each faculty of sense, and keep the heart
Awake to Love and Beauty!
(61–66)

This famously dubious passage about Nature's faithfulness to the right kind of inhabitants has moral as well as ecological resonance. The "wise and pure" are those who dwell in patient contemplation of nature and are rewarded with insights half-imagined, half-existing in the world. Vacant wastes become darling gardens. Since Coleridge is forced by physical injury to stay within a "plot so narrow," his imagination delves deeply into the scene, and he emerges with the microcosmic epiphany that is unavailable to those hikers contemplating a broader range.

Coleridge's glib hope is that Nature rewards "wise and pure" stewardship with pleasant spots and happy thoughts; the remark echoes the moral of "The Rime of the Ancient Mariner" to "love all things both great and small," a sentiment wholly incongruous with the gothic perversity and haunted hollows of that poem (616). Coleridge closes borders around pleasant spots of nature as a form of damage control. In "This Lime-Tree Bower My Prison," the scenery blinds out the reality of Mary Lamb's matricide; in "Rime of the Ancient Mariner," the interscalar love of great and small is a sound bite compulsively repeated by the Mariner, but the mantra never liberates him from the malicious original sin he committed with the crossbow. With his dainty morals, Coleridge doth protest too much. By looking at the microcosm in Wordsworth's and Coleridge's poems, we detect a psychological distinction between the collaborators. William uses the ecological microcosm as an epitome of dwelling and the harmonious powers of the mind in nature. Coleridge's morals attempt to seal out the scarier elements of a darker, more sinister nature that he knows are there, but the seal of morality cracks and the haunting leaks back in.

John Clare, a true native of his verse locales, also contributed to the Romantic cognitive alchemy that transforms nature's small models into universal symbols. Clare composed according to his doctrine of "taste," the susceptibility to the small delicacies of nature that he believed few people possess. Clare's romance with the minutiae of nature appeals to the opening of individual perception, and his favorite subjects are self-contained, such as the series of bird nest poems and the returning theme of a healthy, heterogeneous village commons. Enclosure of the commons is Clare's anathema. Its artificial circumscription of common lands promotes economic privatization over natural boundaries. His elegiac poems about enclosure are simultaneously ethical and epistemological. He protests the transformation of the commons into private lands for aesthetic reasons that also turn out to be ecological grounds for objection. In "Remembrances," he writes,

> By Langley bush I roam but the bush hath left its hill,
> On Cowper green I stray, 'tis a desert strange & chill,
> And spreading Lee Close oak, ere decay had penned its will,
> To the axe of the spoiler and self interest fell a prey,
> And Crossberry Way and old Round Oak's narrow lane
> With its hollow trees like pulpits I shall never see again.
> Enclosure like Buonaparte let not a thing remain,
> It levelled every bush and tree and levelled every hill
> And hung the moles for traitors, though the brook is running still
> It runs a naked stream cold and chill.
> (l.61–70, in *Poems* 124–28)

Certainly, "self interest" hastens the homogenization of Langley as it is transformed from an open, hardly managed commons with a diverse native population into a closed system designed for agricultural economic ends, "All leveled like a desert by the never weary plough" (58). Langley, Swordy Well, and Round Oak commons were models of larger shifts in land use and privatization across England. In this case, the trope of the microcosm is a figure of thought for understanding how each individual act of enclosure implies a cultural shift from ecocentric to egocentric modes of dwelling. Certainly, commons were already highly altered and working landscapes before enclosure, but their identity as communal agrarian spaces with inherent value was dismantled as they were fenced for private ownership. The hung moles, killing traps, felled oaks, exposed brook, and eroded soils are emotional environmental laments, but they also connote ecological concern: landscape deprived of its native flora and fauna is terminally degraded. From a systems perspective, the commons shift from one stable state to another: from half-wild heterogeneity to a highly controlled agricultural system with only a few cash crops. Little input is needed to maintain the first, organic stable state—the villagers' animals are probably the

most important regulators—but the mono-agriculture state requires constant grooming in the forms Clare laments: clearing out wild animals, trees,
and "weeds," and altering the brook for irrigation.

Clare's devotion to nature in the wild makes for an ambivalent view of
the work of natural historians and their role as scientists who both appreciate and exploit nature. His poem "Shadows of Taste" is instructive on
Clare's opinion of science (*Poems* 134–39). His idea of natural taste, which
might be called "close reading" or "susceptibility," is understood through
minutiae. When compared to the avaricious "vulgar hinds" (line 100) whose
only view of beauty in nature is through "self-interest and the thoughts of
gain" (105), Clare's "man of science and of taste" (107) works with a genius
borne on the pleasure principle, and this enthusiasm for nature is largely
redeeming:

> The man of science and of taste,
> Sees wealth far richer in the worthless waste
> Where bits of lichen and a sprig of moss
> With all the raptures of his mind engross
> And bright-winged insects on the flowers of May
> Shine pearls too wealthy to be cast away.
> His joys run riot 'mid each juicy blade
> Of grass where insects revel in the shade,
> And minds of different moods will oft condemn
> His taste as cruel: such the deeds to them,
> While he unconscious gibbets butterflies
> And strangles beetles all to make us wise.
> (107–18)

Clare is drawing a tricky distinction here between the scientist's close study
of his surroundings and his "unconscious" execution of the insects. Wordsworth's line "we murder to dissect" sounds as echo to the strangled beetles,
but Clare's lines are more playful than accusatory in tone, especially since
the scientist possesses Clare's coveted "taste" ("The Tables Turned" 28). The
scientist's gibbeting and strangling is more like a child's rough collection of
natural bounties, in the vein of Wordsworth's "Nutting," than like an image
of the Reign of Terror or other politically charged events. Clare is a lover of
nature in context, partaking of all the wondrous interconnection that makes
the study of ecology necessarily sensitive to holism. The scientist without taste,

it follows, is the one who extracts portions of the system under reductionist conventions:

> But take these several beings from their homes,
> Each beauteous thing a withered thought becomes;
> Association fades, and, like a dream,
> They are but shadows of the things they seem.
> Torn from their homes and happiness they stand
> The poor dull captives of a foreign land.
> (147–52)

The foreign lands of laboratories and cabinets of curiosities are the lamentable reductive counterpart to the study of nature in situ. Clare's subtle allusion to colonial botanical discoveries, the "captives of a foreign land," indicts the practice of collecting in general. In his view, wisdom comes from the existing landscape, not its hacked-out samples. The parts themselves become withered limbs of an organic body not yet understood—collection is too reductive. His final image, a celebration of the wisdom wrought by taste, is concentric:

> Such are the various moods that taste displays,
> Surrounding wisdom in concentring rays
> Where threads of light from one bright focus run
> As day's proud halo circles round the sun.
> (161–64)

These tasteful encounters with nature gather in increasing concentric scales to envelop the macrocosm of the inner solar system. Scalar congruence provides true knowledge, and the parallel between small and large is the only way to reduce natural systems without murdering them. Clare is developing the character of his ideal scientist, who practices what Goethe called a "delicate empiricism," the image of the poet-scientist whose intense attention and empathy work to annihilate the egotistical self. Once ego is diffused, we may gain intimacy with the organism system. Particularly Dorothy Wordsworth, John Clare, and John Keats, the self-proclaimed chameleon poet, were advanced in the philosophy of suppressing ego so as to highlight the ecological muse, which rewards them with a clearer vision of the system.

One last Romantic image merits attention in the context of the micro-cosm. As a counterpoint to the landscape microcosms of the poems discussed above, in "Prometheus Unbound," Percy Shelley imagines a global system of intelligible order and interchange called an "inter-transpicuous orb." In effect, Shelley pushes his metaphysical ideals into a lyrical exposition of machine science based loosely on Ptolemaic theories and on contemporary ideas of matter and electricity (Grabo 141–43). Less attention has been paid to the deep-seeded theme of organicism and biology in this passage, and to the industrial conceit of the machine that is a submerged metaphor sustain-ing motion throughout. Together, the organic richness and mechanical rigor of Shelley's image create an epistemological balance evocative of ecological conceits like the "earth system" or "biosphere," ways in which we have come to comprehend global dynamics in the twenty-first century. Shelley evokes the structure of the atom and the Renaissance cosmos at once (Fraistat 218n36). Shelley's vision unfolds as Panthea, one of the Oceaniades, speaks:

> A sphere, which is as many thousand spheres,
> Solid as crystal, yet though all its mass
> Flow, as through empty space, music and light:
> Ten thousand orbs involving and involved,
> Purple and azure, white, and green, and golden,
> Sphere within sphere; and every space between
> Peopled with unimaginable shapes,
> Such as ghosts dream dwell in the lampless deep,
> Yet each inter-transpicuous, and they whirl
> Over each other with a thousand motions,
> Upon a thousand sightless axles spinning,
> And with the force of self-destroying swiftness,
> Intensely, slowly, solemnly roll on,
> Kindling with mingled sounds, and many tones,
> Intelligible words and music wild.
> With mighty whirl the multitudinous orb
> Grinds the bright brook into an azure mist
> Of elemental subtlety, like light;
> And the wild odour of the forest flowers,
> The music of the living grass and air,
> The emerald light of leaf-entangled beams
> Round its intense yet self-conflicting speed,

Seem kneaded into one aërial mass
Which drowns the sense.
("Prometheus Unbound" 4.238–61)

The dualistic nature of this vision pushes our understanding of the earth sphere beyond landscape into something organized under higher-order, emergent ecological principles perched on layers of biology, chemistry, physics, and mechanics. A machine metaphor of "sightless axles" moving in "self-destroying swiftness," builds a structured momentum that "Grinds the bright brook into an azure mist / Of elemental subtlety." Organic elements appear within this engine like fuel for the earth system "Kindling with mingled sounds," including "The music of the living grass and air." Shelley includes kindling for the other senses: "the wild odour of the forest flowers," the "emerald light of leaf-entangled beams," and the aforementioned "bright brook [ground] into an azure mist." Although the momentum is "self-destroying," it is also eternally creative; the azure mist condenses back to brook form, we imagine, and the "intense yet self-conflicting speed" of the system is swift but self-regulated, like a negative feedback loop.

Though highly abstract, this system, highly organized, perched on the edge of chaos but assuredly under control, projects an inspired vision of ecological complexity. The interlacing of mechanical structure with organic sensuous energy lends the vision a corporeal symmetry that was at the forefront of Romantic scientific ideas of the world. Earth is not merely an exquisite Swiss watch set into motion, but neither is it a morass of biomass of obscure origin and indecipherable purpose. It is an intelligible construct, "inter-transpicuous" from one scale to the next, reducible to components for the purposes of description. Yet the whole of the macrocosmos, conceived as "one aërial mass," inevitably "drowns the sense." To comprehend, one must scale down.

The density of imagery challenges a clear view of Shelley's vision. Calming the fervor of his visionary inspiration, he locates the infant "Spirit of the Earth" asleep, a symbol of our trust in the self-sustaining properties of "Mother" earth (265). But the poetry immediately takes off again by tracing the beams of a star on the Spirit's forehead, a light that penetrates the geological and historical secrets that recent Enlightenment science had revealed. Continuing the conceit of machination, the beams "like spokes of some invisible wheel / [. . .] Make bare the secrets of the earth's deep heart" (274, 279). What they reveal is mineral wealth, "Infinite mines of adamant and gold" (280), but, more important, geological extinctions both human and primordial. The canceling of ancient cycles, Shelley seems to suggest, is essential to the sphere's ongoing creativity. This innermost layer of the orb, where the extinct past records how life has changed through time, anchors ten thousand layers of biomechanical activity. This evidence of death is the essential counterpoint to life, as silence itself gives sound the possibility of meaning. Shelley ends the extended metaphor of the orb with the extinguishing words of a comet/God: "'Be not!' And like my words they were no more" (318). The entire fourth book of "Prometheus Unbound" is regenerative in nature, and this one utopian passage demonstrates the power of a scientifically driven imagination rendering old ideas from Ptolemy, Bacon, Newton, and Milton, and new theories from Erasmus Darwin and Humphry Davy into a Romantic machine of ecological complexity.

Microcosms, as idealizations, summaries, and mechanical constructs, have come to embody some of the most problematic elements of cornucopian views of nature. William Morris would criticize detached visions of picturesque nature as a lamentable inheritance of the "Century of Commerce." In *The Beauty of Life* (1880), Morris politicizes Gilpin's picturesque aesthetic with a simple question: "How can you care about the image of a landscape when you show by your deeds that you don't care for the landscape itself?" (49). It had become clear by the late nineteenth century that the picturesque was an effective way to summarize nature, but that art did not always play the role of environmental preservationist. Along with the general commerce of industrialism, many works of art had come to exploit natural scenes for capital gain, as well-worn tourist paths in England's "wilds" could attest. Even William Wordsworth, ambivalent at best about his following flocks, published a book in 1810 designed to enable a traveling appreciation of Lake District locales. The *Guide to the Lakes* was no Lonely Planet volume, nor was it a vapid rhapsody on picturesque scenery; Wordsworth sought to contribute to a popular genre new ethical arguments about land use, class, economics, and how humans are understood by their modifications and accommodations of natural forms. The microcosm carried a whiff of Romantic idealism into a more practical Victorian age, thus wedding the metaphysical concept of holistic containment with myriad mechanisms that might realize such coherence in ecomaterial terms. What the brain contained, nature reflected, and technology and mechanics might eventually manufacture. The brain-as-world trope can be linked to the biogeochemist Vladimir Vernadsky's concept of the "noosphere," dating from the 1920s. Vernadsky defined it as a new phase of cosmic organization that will come after the geosphere and biosphere phases. Biological life altered the face of the sterile geo-lump that was the original planet earth; Vernadsky drew the corollary of the human brain rising to command the biological world. His mechanism, based on nuclear physics, theorizes how humans could transmute elements to create whatever matter they desired from the basest of inputs. The biological world is subject to wandering evolution through deep time, but the noosphere promotes humans to gods who actively create their desired forms from the willing elements. As a speculative hypothesis, the noosphere is purely metaphysical, but like the most self-absorbed versions of the psychological microcosm, its egocentric tone has implications for the way we manipulate (mind-ipulate) nature.

Romantic-era microcosms afford figurative perspective on the patterns inherent in nature on a larger scale. It is not surprising that poetry would be a favored medium for exploring such interscalar congruence; one of the celebrated features of lyricism is its habit of circling around images, juxtapositions, rhymes, and rhythms in order to gain deep perspective on a subject. Lyric poetry's lens focuses closely on the subject, the lime-tree bower or floating island, before drawing back to place this little world in the larger cosmos.

5

Victorian Microcosms: Domestic Systems

Early Experiments in Modeling

Although the science of ecology did not gain a name or many formal methods until the last decades of the nineteenth century, natural historians and hobbyists were influenced by the microcosm much earlier. Before we look at microcosms in Victorian literature, it will be useful to sample the empirical impulses that began with George Sinclair in the 1820s and continued to appear sporadically until the 1880s, when Stephen Forbes published the first overt microcosm experiment. The Victorian era is particularly rich in natural history because the empire gathered exotic ecologies and domesticated them, while at the same time large swaths of English nature were forced from wild and pastoral to urban industrial conditions. The Great Exhibition of 1851 was a symbolic moment when the Crystal Palace displayed the spoils of empire as a microcosm, and synechdoche, of global dominance. By midcentury, the vogue of the terrarium had fixed the notion of a landscape as a pet project that provided amusement and instruction.

Looking for absolute firsts in history is a daunting task, but Joseph Priestley was probably the first scientist to create a small ecosystem for explicitly experimental purposes. In 1771–72, Priestley conducted a series of tests on the chemical qualities of air by using a mint plant in combination with a candle and a mouse. He found in these two permutations that the plant had the capacity to refresh air that had been befouled by the candle and the animal. His prose is worth quoting:

> Plants, instead of affecting the air in the same manner with animal respiration, reverse the effects of breathing, and tend to keep the atmosphere sweet and wholesome, when it is become noxious, in consequence of animals living and breathing, or dying and putrefying in it. [. . .] I found that a mouse lived perfectly well in that part of the air, in which the sprig of mint had grown, but died the moment it was put into the other part of the same original quantity of air; and which I had kept in the very same exposure, but without any plant growing in it. (193–94)

Priestley's insight into the complementarity of respiration in animal and plant life is one of the most pleasing mutualisms in biology. It shows how photosynthesis is essential to the maintenance of aerobic life on the planet. Enabled by the technology of the bell jar, this microcosm is as basic as it gets, but it shows how far an ecological community can be reduced and still survive in the short term. It eliminates other factors and makes the invisible, oxygen and carbon dioxide, visible by effect.

Priestley had shown that plants and animals are mutualistic, a basic truth in nature borne out by their contrasting life strategies. But what about interspecies relations within the plant kingdom? An English gardener interested in the potential vigor of biodiversity executed what might be called the first formal microcosm experiment a few decades after Priestley's work. Beginning in the early nineteenth century at Woburn Abbey, the seat of the Duke of Bedford, head gardener George Sinclair designed a garden of 242 experimental plots (four feet square) and eighty-five slightly larger beds that surrounded and buffered the plots and housed cover crops like clover and trefoil (Hector and Hooper 639–40). The experimental plots were raised beds made from wood and cast-iron frames, and some had tanks to hold aquatic species. With these 242 minigardens, Sinclair manipulated several variables, including soil chemistry (acidity, density, mineral content) and plant community composition. Sinclair was interested in finding the optimal assortment of grass species to support ruminants in given microclimates, and he pursued this pastoral ideal using an innovative scientific method. He measured individual and community development in the context of chemical analyses of plants and soils. The illustrious Humphry Davy advised him on chemical queries, and results of the earliest soil chemistry in Sinclair's plots emerged as an appendix in Davy's *Elements of Agricultural Chemistry* (1815). The first edition of Sinclair's *Hortus gramineus woburnensis* was published in 1816 as a lavish folio that contained dried specimens of experimental grass species. Later editions in 1824, 1825, 1826, and 1829 continued to update Sinclair's findings and disseminate this new science of long-term ecological research to audiences in England and Germany.

In the heart of the Romantic era, Sinclair was discovering the methods of the ecological microcosm as a way to model and manipulate natural dynamics. Though his scientific methods have been updated in the past two centuries, many ongoing biodiversity experiments use the same method of manipulating experimental plots. These small plots all have the potential to demonstrate ecological dynamics at the much broader circumferences of pasture, ecosystem, and bioregion. In his second edition (1824), Sinclair shares the excitement of the microcosm's scalar intelligence:

> A sufficient combination of superior selected seeds adapted to the soil, and excluding all the inferior and injurious species to which such soil may be obnoxious, promises to furnish the most valuable nursery from which meadows and pastures on similar soils can be multiplied by transporting turf. On the other hand, when a model of superior grass land shall have been created on a small scale from seeds, the process of transplanting may be employed to diffuse it with speedy maturity and undiminished excellence, as occasion may require, over more extensive portions of land. (vii–viii)

Sinclair's business was turf marketing. His motivation was economic—to establish better pasture and more profitable animal husbandry—but his discoveries also served epistemological and ecological ends. His experiments show how to simplify pastures by dynamic modeling rather than reductionist dissection. His findings suggest that there is a scientific grounding in the aesthetic impulse toward biodiversity. A plot with more species generally is more stable, holds more biomass, and therefore supports more grazing. With the self-effacement characteristic of the artist addressing his patron, Sinclair appeals to the value of his work in the nascent ecological sciences: "I flatter myself that I have developed enough to awaken the attention of other observers of nature and lovers of agriculture, and to point their observation to many of the topics which court their inquiry. I shall esteem myself happy, if this work may arouse a spirit of discussion and experiment, in a field which so few have yet trodden, content to be left the hindmost in the march of science, by all those, whose superior zeal and intelligence shall outstrip me in the contest of utility" (v–vi).

A more influential scientific text, Darwin's *Origin of Species*, was indeed touched by Sinclair's work, as it was by other Romantic-era theories from Malthus and Lyell. Alluding to Sinclair's work at Woburn, Darwin writes, "it has been experimentally proved that if a plot of ground be sown with one species of grass, and a similar plot be sown with several distinct genera of grasses, a greater number of plants and a greater weight of dry herbage can thus be raised" (87). From these grounds, Darwin speculates on the power of the isolated microcosm of land or water as a place to demonstrate the ecological bias toward diversity:

> In an extremely small area, especially if freely open to immigration, and where the contest between individual and individual must be severe, we always find a great diversity in its inhabitants. For instance, I found that a piece of turf, three feet by four in size, which had been exposed for many years to exactly the same conditions, supported twenty species of plants, and these belonged to eighteen genera and to eight orders, which shows how much these plants differed from each other. So it is with the plants and insects on small and uniform islets; and so in small ponds of fresh water. [. . .] In the general economy of any land, the more widely and perfectly the animals and plants are diversified for different habits of life, so will a greater number of individuals be capable of there supporting themselves. (87–89)

Postindustrial "grass farmers" who raise high-quality, low-impact animals on small farms have great respect for the structural integrity of a diverse pasture, as Michael Pollan showed in his profile of Polyface Farms. Species coexist in stable numbers only when they occupy different niches—they do not compete for the same limited resources; they coexist. By avoiding lost energy in competition, species-rich grasslands are more productive: they fix more of the sun's energy in biomass and pass that wealth directly on to the grazing animals whose scat fertilizes the terrain. Diverse pastures are also hardier: they can withstand extreme conditions because they have a wider array of genetic tolerances. Like a pentalingual European and a monolingual American in an exotic land, it is more likely that the European will find the words that convey shelter and food. Five grass species have more genes to cover the contingencies of climate than a monoculture does.

Darwin's appeal to "small and uniform islets" and "small ponds of fresh water" shows his empirical modeling instincts at work. He received a rigorous education in island biogeography during the five years of the *Beagle*'s voyage (1832–36). He was rewarded by the opportunity to compare species of finch on the Galapagos archipelago, but he also found striking ecological juxtapositions in isolated oceanic islands. The adaptive radiation of finches across the Galapagos Islands is scientific history's premier lesson in how a series of islands can reveal, in microcosm, evolution at work. But Darwin learned from another island that over the course of a few hundred years colonists can have profound impacts on small, circumscribed environments. During the final leg of the *Beagle*'s voyage, in July 1836, the party landed at St. Helena, an island of 164 square miles in the middle of the South Atlantic. St. Helena is famous as Napoleon's place of forced exile from 1815 to his death in 1821, but before that it was used by a motley series of European sailors starting in the early sixteenth century, when the Portuguese introduced goats and citrus trees.

St. Helena became a permanent British territory in 1834. When the *Beagle* arrived in July 1836, Darwin had grown out of his recalcitrant academic youth to become an energetic and accomplished naturalist, actively considering competing scientific theories of geology, botany, and natural history during his rambles through Oceana. *Beagle* voyagers were steeped in exotic landscapes, but the flora and fauna of St. Helena reminded Darwin uncannily of home. His diary from the voyage, which provided the bulk of his 1839 volume *The Voyage of the Beagle*, reveals his evolving sense of the striking power of invasive species. First, the revelation of the island microcosm: "St. Helena, situated so remote from any continent, in the midst of a great ocean, & possessing an unique Flora,—this little world, within itself,— excites our curiosity" (412).[1] Darwin's impression of the uncanny effect of English landscape in an exotic land has survived in fragments:

> In latitude 16°, and at the trifling elevation of 1,500 feet, it is surprising to behold a vegetation possessing a character decidedly English. The

1. To our great misfortune, this suggestive passage comes directly after a two-page gap in the diary, which, the editor notes, "relates the loss of two inserted pages [of] discussion on the changes in the fauna and flora of the island since the introduction of goats and hogs in 1502." Darwin, *Voyage of the Beagle* 439n58.

hills are crowned with irregular plantations of Scotch firs; and the sloping banks are thickly scattered over with thickets of gorse, covered with its bright yellow flowers. Weeping-willows are common along the course of the rivulets, and the hedges are made of the blackberry, producing its well-known fruit. When we consider that the number of plants now found on the island is 746, and that out of these, 52 alone are native species, the rest being imported, and many of them from England, we see a good reason for this English character in the vegetation. The numerous species which have been so recently introduced can hardly have failed to have destroyed some of the native kinds. I believe there is no accurate account of the state of the vegetation at the period when the island was covered with trees; such would have formed a most curious comparison with its present sterile condition, and limited Flora. Many English plants appear to flourish here better than in their native country. (360–61)

Even to a Victorian patriot, the small world of St. Helena was degraded by the invasive animals and plants that claimed a devastating proportion of the ecological niches. Darwin notes that different species of birds and insects are "very few in number; indeed I believe all the birds have been introduced within late years. Partridges and pheasants are tolerably abundant: the island is much too English, not to be subject to strict game-laws" (364). He makes a survey of the impact of European ungulates on the former forests of St. Helena:

> The fact, that the goats and hogs destroyed all the young trees as they sprung up, and that in the course of time the old ones, which were safe from their attacks, perished from age, seems clearly made out. Goats were introduced in the year 1502; 86 years afterwards, in the time of Cavendish, it is known they were exceedingly numerous. More than a century afterwards, in 1731, when the evil was completed and found irretrievable, an order was issued that all stray animals should be destroyed. (363)

The culling of stray goats in 1731 constitutes an early act of what would later be called restoration ecology. Oceanic islands provided an experimental system by which colonial naturalists learned to discern the series of negative (stabilizing) feedback systems essential to an ecosystem, and how these systems of balance change in unpredictable and severe ways with the introduction of new species. Although these islands became cultural offspring of the imperial power, their regulators, over time, grew wise to the ecological imperatives of the contrasting tropical climate. The adventurous scientists who took posts at the edge of empire were more likely than their counterparts in the metropole to recommend environmental measures such as reforestation. Conservation, by definition, is a conservative practice, but its application in tropical colonies was a radical action that curbed the profits of natural resource extraction.

In the middle of the nineteenth century, Darwin is discussing an ecological phenomenon that had already been observed by colonial governors and wide-ranging seamen for hundreds of years. Richard Grove's *Green Imperialism* elaborates on the importance of colonial exploration beginning in the sixteenth century, particularly the French and English establishments on isolated islands throughout oceanic regions of the equator and Southern Hemisphere. The islands of Mauritius and St. Helena were powerful models of ecology that demonstrated the fine line between ecosystem balance and chaos. Deforestation (for agriculture) caused desiccation and land erosion, and the introduction of nonnative species uprooted endemic communities. Colonialism's most dire impacts were addressed only by a few wise authorities wishing to protect their investments. In the early nineteenth century (just before France lost Mauritius to the British) Governor Decaen instituted restoration regimes, including the culling of animals, fallow time for fields, and reforestation (Grove 153, 257).

Naturalists of the twenty-first century are used to the concept of invasive species: kudzu and bamboo from the East, and ivy, the starling, and Dutch elm disease from Europe. Introduced species often become virulent in new environments because they are freed from the competition and predation of their native communities. Invasive species have certain common characteristics, such as fast growth and reproduction, wide dispersal mechanisms, a broad range of environmental tolerance, and association with humans. The concept of invasiveness shows how ecology, including the evolutionary ecology that is largely Darwin's gift to biology, cannot discount the human factor in any landscape. Ecological imperialism has become an important new way for scholars to discuss the many levels of stress forced on colonial species, human and nonhuman, sentient and vegetable. Islands are the microcosms that first revealed the ill effects of introduced alien species. Ecological microcosms can model the actual biochemical and community-species shifts, and give prognoses on the endpoints of pollution, habitat fragmentation, species loss, and many other new variables imposed by the appetites and wastes of modern industrial society. Microcosms, therefore, can epitomize rapid degradation as well as balanced perfection. The same figure of thought provides a method for investigating dystopian and utopian models of nature, and Darwin found St. Helena to be the former model.

On the British home front, terraria and aquaria that kept exotic species alive in alien landscapes became fashionable in Victorian homes. These domestic microcosms were enabled by the advances of Victorian natural history. In 1842, Nathaniel Ward shed light on sustaining closed plant communities with his work *On the Growth of Plants in Closely Glazed Cases*. Ward's epigraph quotes George Crabbe's epistolary poem "The Poor and Their Dwellings":

> There, fed by food they love, to rankest size,
> Around the dwelling docks and wormwood rise;
> Here the strong mallow strikes her slimy root;
> Here the dull nightshade hangs her deadly fruit;
> On hills of dust the henbane's faded green,
> And penciled flower of sickly scent is seen;
> At the wall's base the fiery nettle springs,
> With fruit globose, and fierce with poisoned stings.
> Above, the growth of many a year is spread,
> The yellow level of the stonecrop's bed;
> In every chink delights the fern to grow,
> With glossy leaf and tawny bloom below.
> (1.290–301)

Crabbe's vision celebrates intense weediness, a cornucopia of wild plants sharing fruit and poison in equal measure. The botanical effusion colonizes impoverished homes, providing spontaneous natural decoration that contrasts the fashion of heavily planned, rectilinear neoclassical formal gardens. The poem recalls Romantic naturalism and the celebration of natural form above artifice. Ward's enthusiasm for botany inspired a series of experiments with container plantings that brought Crabbe's poetic botanical entanglement to life. Terraria balance a scale between art and nature, mechanics and organics.

Living in London in the 1820s, Ward was "surrounded by numerous manufactories and enveloped in their smoke, [so] my plants soon began to decline, and ultimately perished" (25). Ward kept his trials going, in spite of the pollution, with a series of experiments through the 1830s, and gradually he developed the theories and methods needed to assemble a diverse but stable system. He created several large cases to house pseudonatural scenes: these included a fern display called the "Tintern Abbey House" (including a model of the ruin); an "Alpine Case" with high-elevation species; and a "Drawing-Room Case" with exotic palms, ferns, cactus, and aloe. As a literal construction of the picturesque aesthetic, the case substituted for the Claude glass, framing an ideal scene and providing a fashionable live-art alternative to the landscape painting. But Ward was more interested in the scientific insights of his glazed cases, particularly how plants regulate water through evaporation and condensation. The ideal microcosm requires no input of water, minerals, or gases—it only requires solar energy and proper temperature control (glazed cases can become worst-case scenarios of the greenhouse effect). Ward's advances in the science of designing isolated botanical microcosms led to a vogue of glazed-case displays in the mid-nineteenth century that continues to this day. The idea is simple and elegant, and it remains very similar today to the methods Ward originally developed.

Experiments with aquaria began in the late eighteenth century, when keeping a tank was a popular device for naturalists studying aquatic nature in a domestic setting. Before Londoner Robert Warington's 1857 breakthrough of adding live seaweed to a bowl of fish, the water in the tank had to be replaced frequently in order to provide enough oxygen for the animals, at great trouble and expense to the owner. These were perennially imbalanced closed systems, kept alive by the revolution of new, oxygenated water. Once Warington revealed the complementary relationship of nutrient exchange involving the three trophic levels of producers, consumers, and decomposers, aquaria became very popular for their aesthetic and instructional appeal (Beyers and Odum 179). However, the microcosm's analogical power of interscalar modeling in nature remained obscure, in spite of the aquarium and terrarium fashions. Microcosm ecologist Howard Odum writes, "With the exception of a few scientists such as Warington, microcosmic theory in the 1800s was the realm of the philosophers. However, in the first part of the twentieth century, the history of microcosmic thought shifted from philosophy to science, mainly biological science" (181).

The first scientific work that openly used the microcosm as a dual metaphor and empirical method was Stephen Forbes's "The Lake as a Microcosm," published posthumously in 1887. After a series of refusals from
scientific journals, Forbes's lyrical essay eventually saw the light of the press,
and the microcosm quickly became a conceptual aid to the infant science of
ecology. The influence of Darwin is evident in the discursive narrative style
of Forbes's paper. A systems understanding of landscapes conveys organic
sensibility, as though the organization of disparate parts itself constitutes a
higher sense of awareness that can never be appreciated in a machine universe: "Nowhere can one see more clearly illustrated what may be called the
sensibility of such an organic complex,—expressed by the fact that whatever
affects any species belonging to it, must speedily have its influence of some
sort upon the whole assemblage. He will thus be made to see the impossi-
bility of studying any form completely, out of relation to the other forms,—
the necessity for taking a comprehensive survey of the whole as a condition
to a satisfactory understanding of any part" (77). This "comprehensive survey" is particularly taxing to the scientist faced with the task of explaining,
in prose, his hypotheses of the agency of systems without defining an agent.
The discourse of rational science is ill equipped to convey such images of
hierarchy and dependence: it effectively describes experimental design, controls, the outcome of trials, and conclusions. But the lacustrine microcosm,
like Humboldt's *Cosmos* and Darwin's narrative of natural selection, continually appeals to anthropomorphic language to describe its complex agency.
As a result, Forbes's paper reads more like literature than modern science.
He continues:

> First let us endeavor to form the mental picture. To make this more
> graphic and true to the facts, I will describe to you some typical lakes
> among those in which we worked, and will then do what I can (with
> much difficulty and perplexity no doubt, and I fear with no very bril-
> liant success), to furnish you the materials for a picture of the life that
> swims, and creeps, and crawls and burrows and climbs through the
> water, in and on the bottom and among the feathery water plants
> with which large areas of these lakes are filled. (79–80)

The "mental picture" is framed by the boundaries of a small lake. The
microcosm is both literal, in the sense of biosystems, and metaphorical,

because it relies on the reader's imagination to form a "picture of the life that swims." As we will see in the next section, Keats's odes rely on an engaged imagination to delimit their reach. Similarly, Forbes requires that his readers engage their imaginations in order to conceptualize the lacustrine ecosystem set apart from the larger natural world. Forbes's "Lake as a Microcosm" steps between the rich conceptual ground of the microcosm and the literal, cataloguing, and reductive strategies of a nineteenth-century naturalist's inventory.

Forbes's literary instincts are enlivened by his figure of study: "One finds in a single body of water a far more complete and independent equilibrium of organic life and activity than on any equal body of land. It is an islet of older, lower life in the midst of the higher more recent life of the surrounding region. It forms a little world within itself,—a microcosm within which all the elemental forces are at work and the play of life goes on in full, but on so small a scale as to bring it easily within the mental grasp" (77). A microcosmic system creates a small window into nature's workings that human eyes can peek through. Though atavistic, the life-forms within the lake are also denizens of a modern age; though set behind the water's edge, they exchange nutrients and chemicals with the shore; though they are isolated in the "play of life," they are vulnerable to changes on land. Forbes was part of the vanguard of American limnologists who eventually discovered that lakes can be more vulnerable to landscape alteration than adjacent plots of land, and that they can be hotspots of chemicals that become indicator systems for the effects of human pollution. Therefore, the microcosm strategy can demonstrate both the self-sufficient stasis of a semiclosed system and the wild disequilibria that can shift a lacustrine ecosystem into an unrecognizable new order or alternative state. In this characteristic vulnerability, lake microcosms are similar to island microcosms. In Forbes's crucial first vision, though, the lake is an idealized microcosm of life simplified, downsized, and recalcitrant to evolutionary change; easy reading is a virtue that earns it a place among the methods of the new ecology.

To help his audience understand the experimental conceit of a microcosm, Forbes asks that they read his words and attempt to form a "mental picture" of his holistic descriptions. The series becomes an extended *ekphrasis* on the artistic scenes of life under water. As a descriptive supplier, Forbes aims to "furnish you the materials for a picture of life that swims, and creeps, and crawls and burrows and climbs through the water, in and on the bottom and among the feathery water plants with which large areas of these lakes are filled" (79–80). The scientist does not deign to force order on the elements of nature that he supplies to his readers; his lyrical description is meant merely as a "background or setting of the picture of lacustrine life which I have undertaken to give you" (81). Nature's hierarchical holism, its ecological order, must emerge through each individual imagination rather than find obvious demonstration through the eyes of a strictly rational viewer. Forbes writes, "I will next endeavor—not to paint in the picture—for that I have not the artistic skill—but I will confine myself to the humbler and safer task of supplying you the pigments, leaving it to your own constructive imaginations to put them on the canvas" (81). What follows is a passage akin to the boat scene in Shelley's "Ode to the West Wind," except that the language is inlaid with scientific nouns:

> When one sees acres of the shallower water black with water-fowl, and so clogged with weeds that a boat can scarcely be pushed through the mass; when, lifting a handful of the latter he finds them covered with shells and alive with small crustaceans; and then, dragging a towing net for a few minutes, finds it lined with myriads of diatoms and other microscopic Algae, and with multitudes of Entomostraca, he is likely to infer that these waters are everywhere swarming with life, from top to bottom, and shore to shore. (81)

Since the circumference is limited to this small lake and its complement of species, scientific order has a better chance of emerging from the bewildering abundance of life. Ecological systems largely exist on a middle, human, scale, and thus they are directly accessible to our senses, so inferences and hypotheses, borne of imagination, help to simplify the *concordia discors* into a scheme of order amid the chaos.

Forbes uses his lacustrine microcosm to support the tenets of the two theories that were great advances of nineteenth-century biology: biological mutualism and evolution by natural selection. Ecology is a science that resists laws, as a long century of normal science has demonstrated, but these two doctrines of mutualism and evolution bring it a long way toward explaining any ecological system. Forbes explains, "Two ideas are thus seen to be sufficient to explain the order evolved from this seeming chaos; the first that of a general community of interests among all classes of organic beings, and the second that of the beneficent power of natural selection which compels such adjustments of the rates of destruction and of multiplication of the various species as shall best promote this common interest" (87). Very much impressed by the balance that seems inviolable when nature is left alone, Forbes sees his order as akin to a mature climax community that remains in a virtually static state unless disturbed by outside forces. Ecosystems evolve from competitive chaotic clutter toward a sound construct of cosmos, and this is revealed most clearly in model ecosystems like lakes.

Many ecologists today would view economy and balance as quaint ideas derived from Enlightenment economics; Forbes has an antiquated view of ecological organization by twenty-first-century standards. Individuals born in the twentieth century and beyond have little experience of any ecosystem that has not been disturbed by human activities, both primary (clear-cutting of forests, dredging of lakes) and secondary (invasive species, fragmentation and boundary effects). Perhaps a midwestern lake in the nineteenth century felt austere and mature because it seemed truly unaffected by humans, though revisionist history teaches us that the "ecological Indian" had major impacts on the prairies through agriculture, hunting, and setting fires.

For the first empirical ecologists, some initial precepts were essential for teasing cause and relationship from the overabundance of nature at large. Forbes's essay was revolutionary when it was published and is still relevant for several reasons: it reinforces the specific community relations of animals and plants that demarcate ecosystems; it supports Darwinian evolutionary theory as a game both of vicious competition and of mutualism; it reduces the chaos of all nature to a single empirical object that can be approached through ecological close reading.

Forbes concludes his study with an optimistic Darwinian passage reminiscent of the entangled bank trope, which diverts readers' imaginations from the more dire implications of survival of the fittest: "In a system where life is the universal good, but the destruction of life the well-nigh universal occupation, an order has spontaneously risen which constantly tends to maintain life at the highest limit,—a limit far higher, in fact, with respect to both quality and quantity, than would be possible in the absence of this destructive conflict. Is there not, in this reflection, solid ground for a belief in the final beneficence of the laws of organic nature?" (87). These laws, Forbes trusts, confirm humans as a social, cooperative species, but the paradox between destruction and evolutionary creation at the heart of biology remains indissoluble. Where "order has spontaneously risen," we are compelled to believe that the chaos of daily "destruction of life" is a beneficent force in the long run, that evolutionary chaos leads to some hierarchical order, and that the microcosm is a convenient container in which to display this insight. Forbes's legacy influenced the experimental strategies of ecologists throughout the twentieth century, and as the science became more formalized with an increasing number of crises to understand and redress, microcosm experiments became an effective technique for scaling down, simplifying, and accelerating ecosystem dynamics.

In 1935, the British ecologist Arthur Tansley came up with a formal definition for ecosystems. Although they are not literally isolated from surrounding areas, Tansley says that ecosystems are mostly isolated and intradependent in their materials, species, and energy exchange patterns. Tansley was bothered by the muddiness of boundaries in ecology, and yet he recognized that circumscription in nature was often as much a matter of theory and imagination as it was an extant reality that could be studied. To this end, he wrote:

> [Ecosystems] are of the most various kinds and sizes. They form one category of the multitudinous physical systems of the universe, which range from the universe as a whole down to the atom. The whole method of science [. . .] is to isolate systems mentally for the purposes of study, so that the series of *isolates* we make become the actual objects of our study, whether the isolate be a solar system, a planet, a climatic region, a plant or animal community, an individual organism, an organic molecule or an atom. Actually the systems we isolate mentally are not only included as parts of larger ones, but they also overlap, interlock and interact with one another. The isolation is partly artificial, but it is the only possible way in which we can proceed. (299–300)

The boundaries of ecosystem ecology are systems "we isolate mentally." They are half-imagined for the sake of coherent study, and they half-exist as distinct structures of organization in nature. They must be imagined as realities before they can serve as empirical constructs that demonstrate ecological dynamics. Tansley is careful to maintain a materialist position on ecological interrelationships to guard against the trend among ecologists in the early decades of the twentieth century (namely, Fredric Clements) of using quasi-spiritual concepts, including a "super-organism" of the mature climax community (Bowler, *Norton History* 526). Gaia, the global organism that maintains climatic and chemical homeostasis, is the more recent hybrid of scientific and spiritual views of organismal nature.

The tension between materialist and vitalist views of natural systems has no clear resolution, but ecology as a science has had to distance itself from vitalism in order to maintain the formal material terms of scientific theory. This liminal terrain between imagined isolation and literal boundaries exemplifies ecology's struggles with proper methodology. The best hope for simplifying enormously complex natural systems lies in isolating a small portion of nature from all the rest for the purpose of studying it in a kind of intellectual vacuum. Tansley admitted that "the isolation is partly artificial, but it is the only possible way in which we can proceed" (300). The word "artificial" might be replaced by "imaginative," since Tansley is demonstrating how any subject of study that exists in nature will require that its observer define the boundaries of inquiry and make a series of subjective decisions about what is within the lens and what must be excluded for simplicity's sake. The drawing of spheres is more difficult in the middle, ecological, scales of inquiry: solar systems and atoms are conceptually more isolated than climatic regions or animal and plant communities because they are closer to the infinitely large and small. Because of the muddy middle scale, the scale of nature, ecological science has developed methods of imaginative circumscription in order to make empiricism at these middle scales actionable. The microcosm concept performed the much-needed labor of simplification for twentieth-century ecologists, who adopted Forbes's and Tansley's theory that imagination is essential to theorizing a distinct, austere system in nature that can be studied empirically as though it were an isolate.

Literary ecocriticism is often strained between two poles: the writer's aesthetic appreciation of natural beauty and the scientist's reductive investigation into how natural systems work. With the microcosm trope, there emerges a resolution between aesthetics and empiricism. Consider the familiar passage from Forbes in conjunction with Wordsworth's lake scene from the 1805 version of *The Prelude*:

> First let us endeavor to form the *mental picture*. To make this more graphic and true to the facts, I will describe to you some typical lakes among those in which we worked, and will then do what I can [. . .] to furnish you the materials for a *picture* of the life that *swims*, and *creeps*, and *crawls* and *burrows* and *climbs* through the water, in and on the bottom and among the *feathery water plants* with which large areas of these lakes are filled. [. . .] If you will kindly let this suffice for the back-

ground or setting of the picture of lacustrine life which I have undertaken to give you, I will next endeavor—not to *paint in the picture*—for
that I have not the *artistic skill*—but I will confine myself to the humbler and safer task of *supplying you the pigments*, leaving it to your own
constructive imaginations to put them on the *canvas*. (Forbes 79–81,
emphasis added)

As one who hangs down-bending from the side
Of a slow-moving Boat, upon the breast
Of a still water, solacing himself
With such *discoveries as his eye can make*,
Beneath him, in the bottom of the deeps,
Sees many beauteous sights, *weeds, fishes, flowers,*
Grots, pebbles, roots of trees, and fancies more;
Yet often is perplex'd and cannot part
The shadow from the substance, rocks and sky,
Mountains and clouds, from that which is indeed
The region, and the things which there abide
In *their true dwelling*; now is cross'd by gleam
Of his own image, by a sunbeam now,
And motions that are sent he knows not whence,
Impediments that make his task more sweet.
(*The Prelude* 4.247–61, emphasis added)

These two passages are parallel scenes of *ekphrasis*. The description aims
toward both aesthetic description and objective close detailing of the lake
ecosystem's features. The reader's mind is the blank canvas that supports
the pigments of each author's vision. Wordsworth's passage is a metaphor
for autobiography: the lake's diversity is a vehicle for his egocentric vision,
and the empirical eye revels in the details of plant, animal, and mineral in
"their true dwelling." He points to the difficulty of isolating components of
an interconnected scene, and the lake's surface performs doubly as a physical boundary and a mirror. Forbes's empirical description is also a portrait,
but instead of a shadowy doubling of the author, we are asked to imagine
the ecological picturesque—the framed, synergetic system rendered beautiful by its diversity. Though no face is literally reflected on the surface of the
water, Forbes's microcosm still inflects an observer's active construction of a
closed system: the "mental picture" of aqueous flora and fauna relies on the

"constructive imagination" of Forbes's audience. In italic type are some parallels between the two lacustrine images. Poet and naturalist echo each other as they describe in detail (with verbs and nouns) how the underwater world exists as its own interconnected entity, and how it serves as a foil for the observer, whose external position is opposable to the experimental system.

By the turn of the twentieth century, ecology was an established transatlantic discipline, and Stephen Forbes had published the microcosm as an experimental design. The domestication of natural systems, particularly gardens, aquaria, and greenhouses, provided miniature natures analogous to massive national projects like Kew Gardens and the Crystal Palace at the Great Exhibition of 1851. Written in the context of industrialization and landscape transformation, British poetry of the nineteenth century pushed the potential of the microcosm toward the modeling of coherent natural systems. Environmentalists drew ethical arguments from the notion that small acts aggregate to large effects, and from this alignment of macrocosm with microcosm environmentalists drew the mantra "Think globally, act locally."

Space of Refuge: Neo-Romantic Songs and Satires

Nineteenth-century literature contains many of the first reactions to industrial technology's transfiguration of landscape, and, especially by the second half of the century, the ecological microcosm had come to represent symbolic endangered landscapes. Characteristically of environmentalism, poetic reactions tended to be either escapist or incensed, idylls or jeremiads. In the literature of idyllic escapism, microcosms served as food for the imagination battered by the rudeness of urban life and commercial rhetoric. As heterogeneous plots favored by our evolved aesthetic instincts, garden microcosms became refuges within the brick labyrinths of industrial cities. For example, Matthew Arnold's "Lines Written in Kensington Gardens" is a refuge poem where the poet is a charismatic megafauna seeking protected status in his natural habitat, the realm of "rural Pan" (line 24). By contrast, the disturbance of symbolic natural spaces for the sake of economic progress became a source of outrage and whipped up some of the first waves of organized environmentalism in England. As epitomes of larger systems in nature, the endangerment of microcosmic landscapes prompted particularly vitriolic responses. G. M. Hopkins views a felled line of trees as an atrocity against a "sweet especial rural scene" in his "Binsey Poplars" (line 22). Small natural plots held the symbolic potential to represent the fate of all nature under the pressures of population, industry, and economic expansion.

In both idylls and jeremiads, concerns about the fundamental degradation of landscapes fueled Victorian anxieties about modernity. In the world's largest metropolis, Matthew Arnold developed his sense of nature in microcosm using the resource of London's parks. In some ways a revisitation of Coleridge's lime-tree bower, Arnold's "Lines Written in Kensington Gardens" (1852) locates the poet within a minor world of major meaning, where nature holds an ongoing vitality that is usually forgotten in this urban world of getting and spending. Arnold's lamenting voice draws from his perception of alienation among the industrial Victorians. He found his generation caught "Wandering between two worlds, one dead, / The other powerless to be born": the Christian creation story was obsolete, but an acceptable secular or scientific worldview had yet to succeed it ("Stanzas from the Grande Chartreuse" 85–86). In Kensington Gardens, however, the poet finds an enclosed world in the subtle webs of nature that endure at the very center of capitalist production. The gift is psychological calm, a therapeutic reassurance that human concerns are not ubiquitous and lose their egocentric importance by comparison to even the smallest spaces of nature:

> In this lone, open glade I lie,
> Screen'd by deep boughs on either hand;
> And at its end, to stay the eye,
> Those black-crown'd, red-boled pine-trees stand!
> [. . . . .]
>
> Here at my feet what wonders pass,
> What endless, active life is here!
> What blowing daisies, fragrant grass!
> An air-stirr'd forest, fresh and clear.
> [. . . . .]
>
> In the huge world, which roars hard by,
> Be others happy if they can!
> But in my helpless cradle I
> Was breathed on by the rural Pan.
> [. . . . .]

Yet here is peace for ever new!
When I who watch them am away,
Still all things in this glade go through
The changes of their quiet day.
[. . . . .]

Calm soul of all things! make it mine
To feel, amid the city's jar,
That there abides a peace of thine,
Man did not make, and cannot mar.
(1–4, 13–16, 21–24, 29–32, 37–40)

The peace that abides in this insulated metro-pastoral scene reminds Arnold
of mightier forces performing "changes of their quiet day." He benefits from
close attention, the susceptibility of John Clare's "taste." Arnold details the
synergies in the garden in sensual terms: the visual aesthetics that "screen"
and "stay the eye"; the smells of "fragrant grass" and "air-stirr'd forest"; the
sound of Pan's breath, "peace" and "quiet" that force the retreat of the "huge
world, which roars hard by." Even though this space exists as a series of sensual pleasures, Arnold recognizes that Kensington Gardens lives on in his
absence. It lives on ecologically, modeling the animal-vegetable-mineral
material exchanges of all natural systems; and it lives on metaphysically, to
be kept as a personal mental refuge in stressful times. Arnold makes Kensington Gardens a microcosm of a broader symbolic nature that exists outside human concerns, even when the plot is physically in London's center.
The microcosm is coherent, interactive, heterogeneous, whole, and austere,
and Arnold uses it to tap into an environmental sense of hope. Nature
abides even in the epicenter of industrial modernity.

Though London's parks have therapeutic and symbolic value, they are
highly cultivated and managed systems that incorporate nature into industrial and imperial ideologies. Kensington's appearance evolved from the
early eighteenth century to conform to landscape design fashions, and by
the end of the nineteenth century the "black-crowned, red-boled pine-trees"
that Arnold celebrates had been removed to create a longer visual line.
Kensington Gardens is the former private park of Kensington Palace and it
is adjacent to Hyde Park, the site of the Great Exhibition of 1851. Arnold's
poem dates from 1852, and it is notable that the lyric makes no mention of
the global exhibition that had recently bristled at Kensington's gate. This

obvious absence suggests that the "huge world" Arnold invokes is more than just the world of London; it is the actual globe that is turning so rapidly toward the modern future. The "city's jar" no longer represents a singular city, but London now serves as a metonym for all urban industrial landscapes. The Great Exhibition touted the cultural and ecological spoils of the globe by gathering imperial artifacts under the single roof of the Crystal Palace, an architectural wonder of glass and iron designed after organic forms like stems and leaves. It was an early form of biosphere technology. The technoimperial Crystal Palace in Hyde Park and Arnold's garden space in Kensington are inverse ideologies of the microcosm trope. The Crystal Palace draws a sample of global ecoregions under its roof, thus implicitly claiming to represent the whole world and promoting Great Britain as both patron and master of imperial natures. Kensington is an ideal sample of English nature both in its species communities and in their regulation within gardening aesthetic schemes. The Crystal Palace claims imperial dominion, ecological and cultural globalism, and industrial economic priority. Arnold's Kensington Gardens claims local dwelling, ecological and cultural regionalism, and the escape from modern society.

Some Victorian writers were suspicious of the role of the crafting imagination in delimiting landscapes. These paradises offered up by the Romantic imagination were potentially egotistical and solipsistic, and they relied heavily on idyll and jeremiad, the more tiresome modes of environmental discourse. George Eliot made a comical figure out of Romanticism's epistemology of imagination by blending the Romantic egotistical sublime with the Victorian technocratic belief that degraded landscapes could be remediated and sustainability engineered. This curious hybrid of the Romantic brain-paradise and the Victorian machine-genius emerges in George Eliot's "A Minor Prophet," published in 1874 (*Poems of George Eliot* 381–90). It foreshadows Walt Disney's "imagineering"—the profitable conflation of outlandish pleasure-nature with advanced mechanical and computer technologies in Disney theme parks. Eliot's poem satirizes the ideas of a "vegetarian seer" named Elias Baptist Butterworth, who philosophizes "Somewhat too wearisomely" on the uneasy evolution of Christianity into a new technical age (1, 16). Butterworth's physiognomy allows him to absorb the world of "Transatlantic air and modern thought" (23); his hair is brushed back "to show his great capacity—/ A full grain's length at the angle of the brow / Proving him witty [. . .] / his doctrine needs / The testimony of his frontal lobe" (28–30, 32–33). A cognitive analogue to the collective soul, Butterworth inhales the industrial "Thought-atmosphere," made of

> a steam of brains
> In correlated force of raps, as proved
> By motion, heat, and science generally;
> The spectrum, for example, which has shown
> The self-same metals in the sun as here;
> *So* the Thought-atmosphere is everywhere.
> (38–43)

Eliot's funny, sardonic steam-punk indictment of nineteenth-century egotism is measured by a serious concern for the environmental endpoints of its actions. The self-admiring, world-conceiving imagination of the Romantics enabled by the machine-driven agency of the Victorians is a menacing figure to behold. Privileged industrial ideals are stamped upon the body of "our infant Earth":

> When it will be too full of human kind
> To have the room for wilder animals.
> Saith he, Sahara will be populous
> With families of gentlemen retired
> From commerce in more Central Africa,
> Who order coolness as we order coal,
> And have a lobe anterior strong enough
> To think away the sand storms. Science thus
> Will leave no spot on this terraqueous globe
> Unfit to be inhabited by man,
> The chief of animals: all meaner brutes
> Will have been smoked and elbowed out of life.
> (67–78)

As we look back on the long century that has elapsed since its composition, the chilling accuracy of this satirical passage makes Eliot's poem a serious and inspired environmental claim that values "wildness" over faux utopias like golf courses and air-conditioned deserts (the Las Vegas breeze). The protesting voice of Colin Clout identifies how "every change upon this earth / Is bought with sacrifice" (145–46). Technological purchases come at the cost of biological heterogeneity as well as intellectual richness. Remember the idiocy of Wells's Eloi in *The Time Machine*—in the age of evolution, both Eliot and Wells are concerned about the paradox that our clever inventions tend to make us physically lazier and intellectually moribund. Clout takes a position against the eugenic ideal that raged in the wake of evolution by natural selection:

> A clinging flavour penetrates my life—
> My onion is imperfectness: I cleave
> To nature's blunders, evanescent types
> Which sages banish from Utopia.
> (173–76)

Imagineered utopia is uniform and dull. When we design the future around ideals of industrial synergy, physical comfort, and conventional beauty, we substitute a bland and predictable telos for the wild vitality of evolution's chaotic experiments with life. The human mind is an inadequate vessel for containing all the perfection and imperfection it views in nature. Recall Richard Jefferies's prophetic statement late in his career:

> When at last I had disabused my mind of the enormous imposture of a design, an object, and an end, a purpose or a system, I began to see dimly how much more grandeur, beauty, and hope there is in a divine chaos—not chaos in the sense of disorder or confusion but simply the absence of order—than there is in a universe made by pattern. [. . .] [T]his machine-made world and piece of mechanism; what a petty, despicable, microcosmos I had substituted for the reality. [. . .] I look at the sunshine and feel that there is no contracted order: there is divine chaos, and, in it, limitless hope and possibilities. (*Old House at Coate* 163)

By the 1870s, when Eliot and Jefferies were writing, the imposition of mechanical order on nature had become a noticeable threat to the wild mutative vitality of evolution's crucible. Eliot's "steam of brains" and Jefferies's "machine-made world and [. . .] petty, despicable, microcosmos" reveal a moral problem with a culture that believes that technology can continually improve or even replace nature. No matter how complex the technology becomes, the logical deduction is faulty, because the forms of nature evolved through a series of splendid and random serendipities, not through clever bioengineering. Postmodern ecologists of the twenty-first century voice concerns analogous to these nineteenth-century views. Normal ecology often reduces ecosystems to mechanisms we can model using microcosms, which threatens to discount the element of chaotic creativity in landscape evolution. The literary perspective contends that ecological models are highly cultural and subjective, can be vital, creative, and disastrous, and serve as refuge and symbolize wreck. Not strictly limited to the role of controlled reducer of nature, these literary microcosms are symbols that capture the essence of larger ecospheres and the individual experience within the community circle.

6

A microcosm is an ecosystem diorama. It is a key empirical strategy in today's ecology because it provides a model of a larger environment that can be manipulated along any gradient the experimenter chooses. Microcosms have become the major empirical tool for the subdisciplines of ecotoxicology, soil biology, genetic engineering, and systems biology. By using model experimental plots in demarcated fields, terraria, aquaria, and even microscopic plates, today's ecologists are able to predict how ecological systems behave in reaction to disturbance—for example, when atmospheric carbon dioxide reaches 450 parts per million by volume. This physical modeling has been extended to the virtual realm of computer simulations that predict many kinds of ecological impacts, including climate change. Computer modeling is a postmicrocosm analogous to the posthuman android, because the technological body has replaced the biological one. Climate modeling can entertain millions of variables in the attempt to anticipate emergent effects like the nonlinear dynamics of tipping points; this brings microcosm epistemology into play with chaos.

Although Forbes's microcosmic lake, like the microcosm images offered by nineteenth-century poets, takes place outside, in situ, ecologists rapidly brought their microcosms indoors. In a laboratory setting, the models are subject to strict controls, are constantly accessible, and give ecology the prestige of a lab-based science. Ecologists often adopt microcosms as pet projects, like Victorian terraria, and grow fond of their constructions as superorganisms in their own right. Take, for example, the college-level text *Ecological Microcosms* (1993), used as an instruction manual for building and personalizing microcosms according to the owner's interests: "The variety of intricate, small, experimental worlds constructed by various investigators rivals that of nature developed without human hands. [. . .] [W]e hope our readers can share our love of little systems, their mystery, their creativity, their domesticity, their immortality, and the guidance they provide for the larger realms. As living models, microcosms help bridge the details of reality with the abstractions of general systems, revealing the principles of the way all systems work" (Beyers and Odum vii–viii). The enthusiasm of this passage points to both the strengths and the hazards of lab-based microcosm experiments. Microcosms bridge the gap between real and abstract to show how material ecologies can be understood through general principles and theory. But the model's "immortality" figures the microcosmologist as a minor god manipulating her own worlds with an agency that we simply don't possess in the macrocosm (though technological cornucopians and geoengineers are working with zeal). Microcosm models may elucidate some ecological principles, but they assume that manipulation is the proper mode of ecological work, implicitly extending the "pet" analogy from the laboratory terrarium to the wetland and the prairie. James Drake argues that laboratory microcosms afford the clearest possible perspective on the nature of the chaotic forces that direct evolutionary patterns:

> How much of the pattern of nature is the result of stochasticity and simple environmental filtering, and how much is the result of chaotic dynamics, assembly mechanics, and self-organization? This question is fundamental to all aspects of biology, and clearly the first analytical approach must be conducted in the laboratory where tight control is possible. [. . .] The utility and power of microcosm analyses to provide insight into ecological systems is limited only by imagination and creativity. We can think of no questions, from the most applied to the most abstract, to which microcosm analyses cannot be turned and insight gained. The potential for more microcosm studies has steadily increased as new questions concerning biological invasions and introductions, species richness and productivity, global climate change, release of genetically engineered organisms, species extinctions, and other problems facing ecologists are addressed. [. . .] Theory suggests rich dynamics at the cusp of chaos and anti-chaos, dynamics that are

best explored initially under the highly controlled conditions of the laboratory. [. . .] We stress that these systems are but models of the real world and are designed to address specific questions. (Drake, Huxel, and Hewett 675)

Sharon Lawler offers a similar view of the microcosm's necessity in chaos ecology: "Chaos theory is practically impossible to test outside microcosm, because of the large number of generations required. Because chaotic dynamics are possible in many population-dynamic models, empirical work is desperately needed to discover whether real systems are governed by initial conditions and transient dynamics or by equilibrium dynamics" (248). The transition from abstract theory (such as chaos) to specific material demonstration is enabled by the strict control of the artificial environment. Drake and Lawler argue that ecological experiments begin with a spark of imagination, and laboratory microcosms are often the smoothest way of transferring the spark of a hypothesis to a controlled blaze of extensive testing. Since chaos is so sensitive to initial conditions, microcosms enable us to test those seminal states to see whether parallel microcosms tend toward convergence or evolve into wildly different forms, which would support equilibrium and chaos theories, respectively. Microcosms are our keystone for testing evolutionary ecology to understand whether natural systems are mostly coherent, balanced, autonomous, and organized or erratic, patchy, contingent, and chaotic. They are best employed to test specific physical and chemical exchanges within highly simplified versions of natural environments, such as the effect of pollutants like ammonia on specific aquatic organisms (*Daphnia* is a popular indicator genus). A whole complex ecosystem, with its complement of diverse species, abiotic medium, and variable climactic conditions, will not be corralled in a fish tank.

Another advantage beyond the controlled setting is the preservation of the real environment while its model is subjected to thermal and chemical pollution and other forms of stress, like extinction and invasive species. Microcosms provide an alternative to the statistical sampling of species composition in natural environments. Simplicity is a premium amid the complexities of nature, and contained experimentation causes less disturbance than manipulation in situ, which can result in spreading chemicals, dredging, and oversampling. Historically, microcosm experiments tested disturbance by creating it in real environments. E. O. Wilson's theory of island biogeography (with Robert MacArthur) emerged from an island microcosm experiment in the Florida Keys that tested colonization and succession dynamics from a starting point of zero population. Wilson and MacArthur fumigated the mangrove island to ensure that this was a completely animal-free zone. Ecologists should not be conflated with environmentalists: experiments can be heavy in impacts, especially when they aim to model disturbance.

The forms that microcosms take are myriad and diverse. They often involve classic, plot-based analyses, still a key empirical tool for restoration and impact analyses. David Tilman of the University of Minnesota, a prairie ecologist, has maintained a long-term ecological research experiment in the plains north of the Twin Cities for the past two decades. His work aims to evaluate the robustness of plains species communities in relation to their biodiversity. Small plots of plains species in the correct composition—some that are species poor (zero, two, or four species), some that are species robust (sixteen or twenty-four species), and some that are meso-diverse (eight or twelve species)—are cultivated, weeded, and sampled by armies of undergraduate summer interns. Over time, Tilman's widely cited work has corroborated the long-standing theory that environments with more biodiversity are better able to withstand extreme conditions (chaos and contingency), and able to sustain more biomass even though interspecies competition is keen. Basically, polycultures are more stable and productive than monocultures; the long-standing assumption is borne out in the microcosm plots.

Other work by Tilman has shown that high nitrogen levels in soil produce chaotic population fluxes in one species of American grass, rather than simply more grass than in low-nitrogen soils. This study caused, in the words of Richard Leakey and Roger Lewin, "a mixture of consternation and excitement [. . .] [because] from the point of view of biodiversity, chaos is a positive force. [. . .] [E]rratic behavior that flows from the internal dynamics of ecological communities is a force for promoting diversity" (159). The consternation is due to the fact that chaos is much more difficult to understand and analyze than is static balance, yet it appears to be crucial to the health of complex species communities. Microcosms are practical tools easily designed and manipulated, and the only way to tease out this chaotic dynamic is by showing the range of fluxes across differentially diverse microcommunities. The work of Tilman and others further exposes the vulnerability of vast monocultures and "perfect" one-species lawns. Tilman's experimental strategy has a direct lineage back to Woburn Abbey in 1816, where George Sinclair sought the best composition of grasses for English pasture with his phalanx of little plots.

If Tilman is working in Sinclair's tradition of the turf, aquatic ecologists have picked up from the hobbies of aquarium-keeper Robert Warington, who in 1857 learned how to oxygenate the water by adding live seaweed. Tank-based models are especially convenient for testing pollution like oil contamination and changing chemical composition like the acidification of the oceans by dissolved carbon dioxide. In one recent experiment, Norwegian ecologists sought to isolate the effects of pure oil on marine life from the usually co-occurring chemical contaminants from runoff and shipping found in industrial harbors (Vestheim et al.). They used fifteen hundred-liter land-based polypropylene tanks filled with uniform seawater containing phytoplankton, bacteria, and zooplankton. They kept three tanks as controls with seawater only, treated three with pure mineral oil dispersed with acetone, three with the contaminant emamectin (a sea-lice pesticide), three with both emamectin and mineral oil dispersed with acetone, and three with acetone only. They sampled the tanks for marine life and chemical composition on days one through five, the last day of the experiment. The convenience of this artificial microcosm setup is clear: biotic and abiotic factors are controlled and isolated, and the scientists have spatial and temporal mastery over the conditions that would be impossible in an experiment in situ. By constructing their microcosms, they were able to reduce the impossible complexity of the biochemistry in a busy industrial harbor to a few key factors. They critiqued earlier microcosm experiments that used crude oil as the contaminant, because crude oil is itself a mixture of "hundreds to thousands of components," some of which are doubtless toxic to marine life (Vestheim et al. 113). In short, they dissected the marine "body" down as far as ecological reduction is possible.

Vestheim et al. found that the contaminants, either separated or together, did not cause a significant change in nutrient concentration or the biotic community structure compared to the control. Even high levels of contaminants appeared not to cause significant damage to the aquatic communities. To their surprise, they found that the acetone had a positive effect on zooplankton (copepods), which may indicate that it can stimulate photosynthesis at much lower levels than previously thought, so the zooplankton in those three tanks may have enjoyed more nutriment without being suppressed by the contamination. This kind of finding is common: the hypothesis is not supported, but new insight appears that is subject to further testing. Of course, this kind of study provides limited insight because of the extremely short-term exposure, the low number of repetitions (only three for each permutation), and the small populations of aquatic organisms in each tank. Longer-term exposure might yield strikingly different results, and the chronic degradation of certain environments is the rationale for funding so many long-term microcosm studies. Still, this kind of quick science produces informative sketches of the complex dynamics in real environments, and it has a practical advantage of sharp biochemical parameters and physical perimeters: the walls of the tanks. The scientists can always open the plughole on Friday, and refill the tanks to test new hypotheses on Monday.

The microcosms described above are classic designs resembling their centuries-old counterparts. Since they are on a human scale, measured by meters and liters, many scientists would call them *meso*-cosms. Twenty-first-century technological innovations can transform modeling into the truly micro scale that is highly abstract and technological. For example, Wittebolle et al. used bacteria populations dwelling in wells in microplates at a hyperurban concentration of ten million per milliliter to test the effects of environmental stressors like pH change, salinity, and temperature in relation to species richness and evenness (richness is the number of species in a community; evenness is the relative abundance of individuals of different species). The scale of the organism used in this experiment in a dish allows for much higher n values, quick results due to bacteria's rapid life cycle, and the convenience of microcontainment. Like many experiments in microbiology, the work can be done from a lab bench, and in this particular case it seems that more computers and robots were involved than graduate students. Some of the tools that allowed for greater convenience and precision were flow cytometry, robot pipetters, ultracold freezers, and "spectrophotometric microplate readers" (Naeem 580). Yet it is still an *ecological* experiment with results that can be extrapolated onto the meso-cosm in which we dwell. The scientists found that not only species richness, as expected, but also species evenness is important to ecosystem resilience in the face of environmental stressors. The more uneven the species composition in communities, the more their functioning and services decline as conditions change (Wittebolle et al. 623). This microscopic microcosm was able to show how a vaguely understood factor like species evenness can have dramatic effects on the fates of ecosystems, and so the concept deserves much more attention from disturbance ecologists. Though unrecognizable to the founding ecologists who worked on human scales in the field, this kind of microcosm still uses live organisms and real chemical conditions, unlike the computer models discussed later in this chapter.

So scientific microcosms can be more or less "natural" (outside or inside, evolved or assembled) and more or less "truthful" about how nature operates on larger scales. Natural and truthful are not always in step in microcosms. A model natural system may be anecdotal or exceptional rather than truly representative of other systems. An artificial microcosm can be assembled to animate the most broad-based applications to a variety of real-life cases. The level of analysis is also essential: a koi pond in a suburban yard is not instructive about species composition in freshwater lakes, but it does demonstrate how nutrients cycle through closed systems.

Critics of abstract ecological modeling contend that microcosms paint erroneous simplicity over what is a naturally complex set of relationships, and that the practice removes ecologists from the study of real environments. Limnologist Stephen Carpenter worries about the demise of a practical education in nature-based ecology: "Who will train the ecologists needed for field science? It is irresponsible for academic ecology to produce larval microcosmologists by canalizing graduate students into careers of small-scale experimentation. There is cognitive danger that the microcosm (rather than the ecological system) will become the object of study, leading to needless confusion as results are over interpreted and over extended. As ecology becomes more and more a science done indoors by urbanites, there is significant risk of losing our sense of context" (679). Working in the older tradition of Stephen Forbes, who found his microcosm in nature, Carpenter reminds us that modeling can paradoxically obstruct our perception of the natural system we mean to restore and protect. Beyers and Odum reveal this hazard as they celebrate their "love of little systems, their mystery, their creativity, their domesticity, their immortality" (viii). In an era when virtual lives are replacing actual ones, attention shifts from the endangered species in a degraded environment to the mechanical songbird on the shelf.

The microcosm poems of previous chapters are trained on real natural systems and preserve the character of the subject while inscribing its complexity in image and prosody. Romantic organicism preferred wild systems to pet ones, and the newfound industrial-era value of nature drew the poet back outside to find representative subjects in context. Romantics were reacting to the artificial subjects and prosody borne of eighteenth-century farce, sentiment, and comedies of manners (George Crabbe, Oliver Goldsmith, and

Thomas Gray provide important exceptions to this Restoration-era vogue of practiced artifice). Consider an analogy between prosody and ecology: critics of indoor ecological modeling are concerned that the natural system in situ is an ode, a fast-fading violet covered up in leaves, and today's students of ecology are only learning artificial techniques that harness together heroic couplets and contrived plots. The former is a deer-filled mixed-grass meadow, the latter, a video game of countryside hunting in which the player who shoots the biggest buck wins. Carpenter's aversion to "urbanite ecology" shows how the evolution of ecology has pushed its empirical methods from natural contexts to artificial models of nature. The heroic history of early ecology celebrates an athletic, sinewy, mucky science that has been geeked up by lab and computer-based microcosm experiments. By keeping attention on actual nature, rather than on proxies of natural systems or samples removed from their natural context, Carpenter and his naturalist colleagues are revisiting the protests of John Clare, who defines "taste" in scientific practice as inherently contextual:

> But take these several beings from their homes,
> Each beauteous thing a withered thought becomes,
> Association fades and like a dream
> They are but shadows of the things they seem;
> Torn from their homes and happiness they stand
> The poor dull captives of a foreign land.
> ("Shadows of Taste" 147–52)

For better or worse, since Forbes's time scientific microcosms have often moved indoors to take advantage of highly controlled conditions. Inside the lab, there is no need to hire armies of interns to eliminate drift-in seedlings. Dwelling in a sensuous ecosystem, feeling the weather as it happens across the seasons, parsing self from environment or allowing them to intermingle—these modes of exposure and intimacy are sacrificed in laboratory microcosms. Instead, peering through glass at a captive model of an ecosystem, the scientist has a clear sense of object, control, and rigid edges.

Furthering this trend of abstraction, lab work has given way to computer modeling. In the past three decades, modeling techniques in all the sciences have moved from the field and lab into the *virtual* world, a revolutionary step that affords exponentially higher levels of variation and manipulation. Global climate models (GCMs) attempt to approximate the actual complexity of the biosphere they model—something that could never be done in a fish tank. These techniques are particularly valuable in the sciences that are nearly impossible to domesticate, such as climatology, which often entertains millions of variables in its analyses. The teams that design and maintain climate change models demonstrate, by their frequent points of disagreement, how many strategies could potentially result in an accurate model of the earth's emergent fate over the next few hundred years.

GCMs are not empirical or falsifiable or objective, so they are not scientific in the classic sense; they are technological. They provide glimpses of the future that we must consider and revise using retrograde evidence, such as the paleoclimate ice cores that speak of earth's remote atmospheric past. This is not to trivialize their importance. The art of global climate modeling is adolescent, and falsifiable models are possible to construct by running them in retrograde. Technicians can run a model of what happened to the climate in the twentieth century, for example, and see if they produce an accurate portrait of the year 2000. One reason that the models vary so widely is that positive feedback loops, tipping points, and other nonlinear dynamics have the power rapidly to alter the way things may come to be. Another source of error is our uncertainty about the ultimate effects of certain climate phenomena—for example, the role that clouds play in either cooling the surface by reflecting solar energy back into space or warming the surface by increasing the amount infrared radiation absorbed (Soden and Held 3354–60).

GCMs hardly need avatars to argue for their necessity. When it comes to predicting climate change for centuries to come, they are the only card we have to play. These models are distinct from classic, material microcosms that focus on communities and ecosystems, because GCMs are predictive on a global scale. GCMs are subject to the same hazards of solipsism as laboratory microcosms. The model run by the computer is itself so complex and compelling that the state of the real environment may vanish behind the technician's need to tinker and perfect the simulacrum. The virtue of computer models, including GCMs, is also a hazard: their complexity helps predict global outcomes, but the way they arrive at these outcomes is only comprehensible to a small cohort of experts, often only the model builders themselves. By aiming to emulate natural complexity through technological complexity, GCMs can be as difficult to understand as the global system they emulate (Dodson 19). Environmental philosopher Lucien Boia assigns heavy responsibility for the spread of "the most alarmist scenarios" to our dependence on models, which easily conflate existing reality with the virtual predictions that come out of extreme, parameter-driven systems. But with proper caution, he allows for their necessity, as "reality is too vast, too complex and chaotic to approach directly. [. . .] They are extremely useful as long as we remember that they are not the real thing: they are methodological fictions" (177).

Of course, we have few ways of empirically, physically demonstrating the superiority of one methodological fiction over another because the essence of prediction is to prognosticate, and then wait and see. Computer modelers represent a new branch in the genre of science fiction. Will the future look more like *Oryx and Crake* or *Walden Two*? Does the Hadley Centre in England or the National Center for Atmospheric Research in the United States have a more accurate climate model? It depends a lot on where you look: utopian temperateness for Canada, Siberia, and Greenland (if only it had topsoil); dystopian chaos in the flooding of Florida and Micronesia and desertification in Botswana and Greece. Part of the computer modeling controversy is philosophical, as Amy Dalmedico has pointed out: "Modeling practices, always pulled between abstraction and application, now found themselves subjected to another set of contradictory forces: should they be first and foremost predictive and operational, or cognitive and explanatory?" (126). Ideally, a model that is explanatory, able to accurately recapitulate earth's climate in the past, would serve as a precursor to a model that is predictive, able to suggest one or several future scenarios according to the manipulation of variables. But the model must traverse the shadowy chasms of emergent effects, tipping points, and similar chaotic behavior. Few experts believe that climate change is a linear phenomenon, and much consternation surrounds the potentially rapid tempo of glacial melting, the oceans turning from carbon sinks to carbon flows, increasing industrial emissions, and collapsing ecosystems.

Philosopher of science Mary S. Morgan cites the epistemological doubt that plagues the practice of virtual modeling because the "medium of representation found in mathematical models differs so much from the real geological or weather events they are taken to represent. [. . .] Even for believers, the inference power of experiments with such [virtual] representations is necessarily weaker compared to those from experiments with representative whole-life models" (270). As we have seen, microcosm science has evolved over 150 years from nature-based sampled systems to lab-based representative systems, and finally to computer-based virtual systems that are the furthest removed from physical nature but may well be the best equipped to simulate natural complexity on a global scale. Morgan has recently developed her ideas about a science that is not dependent on laws but uses representative cases, model systems, and exemplary narratives to move inquiry forward. The last strategy, an exemplary narrative that "converts our experiments in life into experiences," involves narratives that "are taken to say something about a wider set of particular cases or situations than the ones from which they grew. This wider relevance indicates how such objects gain the autonomy to function more broadly as instruments of inquiry" (269, 273). As a complement to modeling, literary scholars have the expertise to develop the potential of narrative in ecology.

Since the time of the ancient philosophers, we have clarified natural structures by imagining interscalar parallels from the whole planet to the human body and down to the atom. Medium-sized structures like the ecosystem are particularly tricky to isolate and model, so microcosms have found their métier in experimental ecology, where conditions can be closely controlled. The trend of control and abstraction gains entirely new dimensions as models proceed from the material world to the virtual world of computer processors, which manipulate millions of variables and rely heavily on the success of the model design in their ability to forecast the outcomes of ecological impacts. However, there is danger in these strict controls, both material and virtual; the systems risk becoming decontextualized and oversimplified the further they are removed and abstracted from their sister systems in nature.

Microcosms in poetry are not empirical, and they are often idealized portraits of the subject. It is a rare scientist who would consider them relevant to her experiments. What is enticing about them, beyond the aesthetics of literary enjoyment, is that poetic microcosms empower an ecological imagination that is richly holistic, constrained only by the enabling structures of meter and rhyme, and open to the surprising and poignant ruptures fundamental to a nonlinear view of change in nature. Where microcosms in ecological science are reductions of the natural system they model, effectively flattening a complex entity into a single layer of analysis, poetic microcosms use circumscription to focus and draw out the depth of subject. They are a form of poetic meditation that brings a specific natural environment into the foreground of analysis using the tools of prosody, image, and allusion. To further develop this hypothesis, which elevates the literary imagination into a dialogue with ecological science, let us consider the poetry of John Keats.

PART 3

Keats and Ecology: A Case Study

7

While case studies are most commonly used in social science, they can be applied as well to interdisciplinary literary analysis, as in the present study. In literary studies, exemplary narratives often show how works by individual authors demonstrate the multifaceted spirit of their age. Case studies are useful for generating and testing hypotheses about similar historical contexts. The brief and intense career of John Keats provides an exemplary case study of discourses of nature, and more specifically the conceptions of chaos and microcosm in Romantic poetry of the early industrial era. Detailed data collected and analyzed from this exemplary poet reveal underlying principles that apply more generally. Keats's period, the latter half of the Romantic era, was a crucial time in modern industrial society. Classic narratives of eternal, balanced nature came into question with evidence of geological catastrophe, biological evolution, natural disasters, and industrial pollution. Writers sought to recover ecological coherence, especially in formal lyrical works that recall the aesthetic of an ordered universe. Since Keats was writing a full half century before the science of ecology established its most basic methods, we can look to his work as a forerunner to ecological thinking, including the development of ecological modeling. Keats has his own interdisciplinary aspect. His poems are the work of a writer deeply interested in the science of his own time. Contemporary notions of evolution, catastrophe, electricity, anatomy, and disease influence his representations of nature.

During a period of concentrated productivity in 1818–19, Keats developed an intriguing tension between narratives of natural catastrophe and models of natural synergy. The vying paradigms of chaos and balance play out in Keats's subjects, scenes, and prosody, from Saturn to the nightingale, Tartarus to the plum-tree bower, and blank verse narrative to lyric ode. Beyond offering detailed landscapes and subjects immersed in a sensuous natural world, Keats's poems notably contain at least two ecological visions anticipating science. The first is the theme of contingency in natural history, as found in his epic fragment *Hyperion*. Scientific readings of this text have typically emphasized Keats's presentation of a coherent succession from Titans to Olympians in order to establish an early evolutionary theory and to reinforce the Romantic theme of political revolution. Thus when Oceanus invokes the principle of "beauty" to govern the defeat of one ruling class by another as a "law of nature," Hermione de Almeida argues that *Hyperion* is teleological and evolutionary (249–51). An alternative perspective emerges with a reading that attends to the environmental disturbance that catalyzes the fall of the Titans. This narrative catastrophism can be traced to Keats's knowledge of French geology and his belief that "chance" rather than classic religious Providence governs life histories.

A second innovation in ecological vision involves Keats's translation of natural spaces into intelligible systems. His lyrics preview a conceptual strategy for isolating and simplifying parts of nature for the purpose of study and anticipate the ecological microcosm introduced to science later in the nineteenth century. Keats's odes also anticipate the theory of ecosystem that Arthur Tansley would define in 1935 as part material, part mental isolate. With Tansley, an ecosystem came to mean a natural unit or area consisting of biotic and abiotic factors operating in synergy; their isolation, however, is at least party conceptual. By reading several of Keats's prominent lyrics as epistemological advances in the science of ecology, these aesthetic successes claim a more utilitarian status than what accrues to pure art for the sake of beauty. From early instances in the fragmented narrative of *Hyperion* to the systemic wholeness of the celebrated odes, Keats reveals a dual relationship between a narrative of chaos, chance, and evolution, on one hand, and a lyrical sense of cosmos, coherence, and stasis, on the other.

Keats is an appealing figure for scientific literary criticism because of his medical education. Hints of Keatsian physiology are abundant in his verse, perhaps most poignantly in his embodiments of health and sickness, joy and melancholy. His biography supports book-length studies of this intellectual synthesis: the born poet is forced by financial necessity into medicine during a time when science and literature were equally relevant epistemologies with philosophical crossing over.[1] Organic form was the aesthetic driving early evolutionary ideas like Erasmus Darwin's, the German Romantics' *Naturphilosophie*, and the artistic instinct to counter Enlightenment mechanism with organic growth and development. Keats's letters inhabit a liminal space between scientific and humanist speculations, and his ambivalence about his profession only heightens the potential for interdisciplinary insight in his poetry. The subdividing of Keats's brief career into distinct phases of composition style is common in his biographies, which show his hyperconsciousness of failure and his sense of embarrassment that leads to his disavowal of earlier modes of thought and writing (see Ricks; Chandler 395). While Keats lore often emphasizes the formative role of the cutting critical reviews of "Endymion," the perspective of periodic artistic evolution from one mode of creativity (epic narrative) to a contrasting mode (confessional lyric) provides an enticing point and counterpoint in the architecture of his prosody in the years 1818 and 1819. Keats's ideas about the individual agonist in the natural world evolved during these years through his open mediation between chaos and cosmos. A purely ecological reading of Keats that relies on the aesthetics of his imagery might appear as a clear departure from the politicized Keats and a return to the old New Critical model of organicism.[2] These more recent critiques claim that Keats was conscious that his narratives of political and ideological maturation are relevant to the agitations of the 1810s, and critics weigh this social engagement equally with his evident growth in style and prosodic mastery. Keats certainly deserves such attention, as the biting criticism levied at his early work was often politically calculated to embarrass the Cockney poet and exclude the lower classes from literary success.

Instead of sparking up the older interpretation of the poetic genius living in a sociopolitical vacuum, this ecological reading engages Keats in his context by appealing to his knowledge of science, which was one of the major forces providing thrust to political reform in the 1810s. Keats was forced by his social standing into medical studies, and though he vacillated between

1. Goellnicht's Poet-Physician, de Almeida's Romantic Medicine, Richardson's *British Romanticism and the Science of the Mind*, Bewell's *Romanticism and Colonial Disease*, and Allard's *Romanticism, Medicine, and the Poet's Body* consider Keats and his poetry in a medical context.

2. For work that offers rich political contexts, see McGann on the Peterloo Massacre (*Beauty of Inflections* 58) and Cox on Leigh Hunt's influence (85).

the practicality of a medical education and the gravitational pull of a literary one, many of his poetic works are doubtless enriched by his knowledge of geology, chemistry, and biology. Keats's context in Regency England is invigorated by his biological conceptual innovations. As an innovator, he participates in the same cause of reform as the Enlightenment "political" scientists so well known to this era, Joseph Priestley, Erasmus Darwin, Humphry Davy, and young Percy Shelley, who conducted chemistry experiments in his rooms at Eton and Oxford. Criticism that touts the physiological Keats, the blushing Keats, the gustatory Keats, is part of the same project of reclaiming the poet from the idealisms of New Criticism and planting him firmly in his environment, and our own (see, respectively, Allard, Ricks, and Gigante).

Keats's experiments in verse are interdisciplinary, effecting recombination between canonical literature (Chaucer, Spencer, Shakespeare, and Milton especially) and contemporary scientific understandings of the world. The revolutionary chaos envisioned in *Hyperion*, and the several microcosms described in his lyrical odes, are attempts to create systems of understanding the world using the complementary tools of empiricism and humanism. The applicability of Keats's verse to early nineteenth-century concepts of nature both installs his poems among contemporary scientific-industrial innovations of the early nineteenth century and links his intuitive ideas about organization of the natural world to emergent ecological theory. *Hyperion* was probably influenced by Keats's reading of Buffon, the French natural historian who wrote voluminously about environmental catastrophe and identified some major tenets of evolutionary theory a full century before *On the Origin of Species* (Bowler, *Evolution* 75). Buffon's encyclopedic *Histoire naturelle* (1749–88) was translated into English, and Keats read the copy in the Guy's Hospital library. Its vacillations on crucial topics like species definition and fixity versus mutability make a clear interpretation of modern evolutionary theory difficult. However, Buffon's secular materialism, his interest in geological upheaval, and his struggles to accommodate these ideas to the religious hegemony of predetermined design in nature make Keats's reading of Buffon germane to this study of ecology. Keats refers to Buffon three times in his letters, and Buffon's *Natural History* was on Keats's mind during the spring of 1818 (Atkinson 341). Though the evidence is circumstantial, the imagery in *Hyperion* resembles Buffon's speculations on the appearance of ancient earth and its tumultuous natural history, a comparison I will return to later in the chapter.

Keats's letters can fruitfully be read in dialogue with his verse, permitting a thread to be drawn between his prose-based distillation of life experience and his infusion of these insights in poetry. Before involving his poetical works, it will be useful to discuss a few turns of thought that Keats developed in a long letter of 1819 to his brother George, who had ventured into the wilds of the young United States. Written from February to May of that crucial year of intellectual development, the letter lies at the juncture between Keats's abandoning the epic *Hyperion* and his rejuvenating his work with the ode form. Late in 1818 Keats lost his youngest brother, Tom, to consumption, and had himself been struggling with chronic ill health since his early return from a walking tour of Scotland in the summer of 1818. Based on these misfortunes, among other, financial hardships, Keats dropped his tepid faith in Providence to adopt a new respect for the power of chance, a word that frequently appears in his letters and poetry after the death of "poor Tom" (his personalized allusion to *King Lear*). He writes to George, recalling an opening image from *Hyperion*, "Circumstances are like Clouds continually gathering and bursting. While we are laughing the seed of some trouble is put into the wide arable land of events. While we are laughing it sprouts, it grows and suddenly bears a poison fruit which we must pluck" (*Selected Letters* 270–71). This agricultural allegory plays out on the stage of environmental contingency. Keats's metaphor involves both the unpredictable, violent weather events of storm clouds "gathering and bursting" as well as the pernicious seeds that germinate as a result. Circumstance, then, involves the contingencies both of weather and of the presence of the organic conditions that foster the unforeseen: the seed to the wide arable land to the poison fruit that we "must pluck." Here, the individual will wields some degree of agency among the chances of life: "There is an ellectric [*sic*] fire in human nature tending to purify, so that among these human creatures there is continually some birth of new heroism" (271).

Keats's chemical opposition between ill circumstance (poison) and fiery heroism (purification) is the alchemy of epic agony, and his fragment *Hyperion* grasps after different ways to empower the Titans, the obsolete gods pondering war with the new Olympians. His "ellectric fire" of purification also has physiological origins in metabolism, where a fire in the belly destroys other life to gain the energy for order, or negentropy, that sustains the human body. In Buffon's view, metabolism and reproduction are analogous across scales, and destruction predicates survival.

Animals seem to participate in the qualities of flame; their internal heat is a kind of fire; therefore, after fire, animals are the greatest destroyers, and they assimilate and convert into their own substance every matter which may serve them for food: but although these two causes of destruction are very considerable, and their effects perpetually incline to the annihilation of organized beings, the cause of reproduction is infinitely more powerful and active; she seems to borrow, even from destruction itself, means to multiply, since assimilation, which is one cause of death, is, at the same time, a necessary means of producing life. (180)

We murder to digest. Keats's "new heroism" is kindled by the fuel of our quarry, which we assimilate into our own fleeting order of the spirit-body. Even the "poison fruit" that circumstance will have us pluck may be purified by the fire of our life and assimilated into the stronger survivor. If animals survive the poisonous elements using their inner fire, they have a chance to reproduce and leave the world to a future generation: dismantle the old, strip it down to its elements, to assemble the new. Here, Keats inherits from Buffon not only a sense of chaotic contingency but also the closed loop of elements that cycle through bodies and across generations.

Not all fires burn with the same fervor, and Keats's letters stir the metaphorical coals to recover his spirits. Keats's faith in Providence was squeezed out under the weight of his younger brother Tom's illness in 1818. His letters start to resist an orthodox rehearsal of Anglican teachings to explore a more independent-minded interest in the power of contingency (Kerrigan 289). Keats's concept of chance comes to inform his theory of an individual's fitness in the inclement natural world. Unlike millenarian philosophers like William Godwin, Keats does not believe in the perfectibility of nature through idealistic philosophy or technological manipulation. Natural adversity is an essential component in his scheme of soul making. Again, Keats uses Lear as a touchstone for true adversity, though he slightly misquotes the king's labeling of disguised Edgar as a "poor, bare, forked animal":

The whole appears to resolve into this; that Man is originally "a poor forked creature" subject to the same mischances as the beasts of the forest, destined to hardships and disquietude of some kind or other. If he improves by degrees his bodily accommodations and comforts,

THE LITERARY EMPIRICIST 161

at each stage, at each accent there are waiting for him a fresh set of annoyances. [. . .] In truth I do not at all believe in this sort of perfectibility. The nature of the world will not admit of it; the inhabitants of the world will correspond to itself. Let the fish philosophise the ice away from the Rivers in winter time and they shall be at continual play in the tepid delight of summer. [. . .] Suppose a rose to have sensation; it blooms on a beautiful morning, it enjoys itself. But there comes a cold wind, a hot sun; it can not escape it, it cannot destroy its annoyances. They are as native to the world as itself. No more can man be happy in spite, the worldly elements will prey upon his nature. (*Selected Letters* 289–90)

Success in life is the growth into individual potential. But this potential is not absolute or predetermined or part of a deistic design; it is defined by the antagonistic elements of nature and happenstance. Life experience can be killing, or it can be ennobling. Using this modern, post-Providence convention of history as chance, Keats has opened an avenue for his materialist spiritualism, which he describes using a scientific trope that had recently been envisioned by John Dalton in 1808, the ultimate chemical unit of the atom. Thinking of the natural world in terms of atomies that organize into greater molecular assemblies advanced chemical and biophysical theory and became part of scientific discourse during the Enlightenment. The biological equivalent of the atom, the cell, was identified by a precocious Robert Hooke in 1665, but cell theory in biology did not develop until 1839 with the work of Schleiden and Schwann. In the 1781 English translation of *Natural History* by William Smellie, Keats read Buffon's description of "living organic particles" that were the building blocks of organic life: "It appears [. . .] that there really exists in nature [an infinity] of small organized beings [. . .] that these small organized beings are composed of living organic particles; that the assemblage of these particles forms an animal or plant, and consequently that reproduction, or generation, is only a change of form made by the addition of these resembling parts alone, and that death or dissolution is nothing more than a separation of the same particles" (Buffon 173). Philosophies of the gestalt and organicism are clearly allied with scientific materialism in the sense that living forms can be reduced to universal atoms or particles, but the scientific clarity given by the act of reduction sacrifices the emergent organization and greater potential of the indissoluble whole. Keats is interested in preserving the clarity of the atomic vision while giving it the behavior of the complex organism. In his spiritual but antireligious philosophy of individual growth, the atom of perception is animated by a spark of intelligence. This charged, perceptive atom works toward the goal of the soul through a struggle to survive the unpredictable circumstances of life, both natural and social.

> Call the world if you Please "The vale of Soul-making." Then you will find out the use of the world. [. . .] I say *"Soul-making,"* Soul as distinguished from an Intelligence. There may be intelligences of sparks of the divinity in millions, but they are not Souls till they acquire identities, till each one is personally itself. Intelligences are atoms of perception; they know and they see and they are pure, in short they are God. How then are Souls to be made? How then are these sparks which are God to have identity given them so as ever to possess a bliss peculiar to each one's individual existence? How, but by the medium of a world like this? This point I sincerely wish to consider because I think it a grander system of salvation than the chrystain [*sic*] religion, or rather it is a system of Spirit creation. This is effected by three grand materials acting the one upon the other for a series of years. These three Materials are the *Intelligence,* the *human heart* (as distinguished from intelligence or Mind) and the *World* or *Elemental space* suited for the proper action of *Mind and Heart* on each other for the purpose of forming the *Soul* or *Intelligence destined to possess the sense of Identity.* I can scarcely express what I but dimly perceive, and yet I think I perceive it. [. . .] Do you not see how necessary a World of Pains and troubles is to school an Intelligence and make it a soul? [. . .] This appears to me a faint sketch of a system of Salvation which does not affront our reason and humanity. (*Selected Letters* 290–91)

The atom (matter) is intelligent (energized) and embedded in a "World of Pains" (ecology). Keats's sketch of life history is a metaphysical anticipation of evolution by natural selection. There is a struggle for existence that leads to differential survival, and each survivor's victory is to become "personally itself," a sketch of genetic and behavioral success. Though mischance and circumstance are lamentable for the woes they bring, natural extremes are essential to a life experience of depth and quality. Each individual's struggles articulate, through scars and privations, an intellect guided by a soul of survived experience. Finding an identity is the heart of salvation; it is a personal acquisition achieved by those possessing enough pith and energy to survive the chance-driven world of elemental space. Though there is some heroism implied in surviving the world, Keats is as interested in exposing how fortune and dumb luck factor into survival—a precocious notion of evolution indeed. These ideas have evolved from Keats's formulation of the mind as a mansion whose corridors and apartments are brightened by trial and experience, which he articulated in a letter to Reynolds from a year earlier (3 May 1818). The passage quoted above goes beyond cognitive architecture. Keats's "vale of Soul-making" is a dwelling place in which each atom of individuality comes to reckon with its existence in chaotic nature. His ode spaces can be inhabited as temporary refuges from the hazards of a contingent natural world. The map by which Keats navigates his system of salvation shows two geographical features: the hazardous road of epic narrative, and the closed bowers of the lyrical ode.

8

Hyperion: The Chaos of Tartarus

The blank verse narrative *Hyperion* is Keats's versification of the "vale of Soul-making." This epic fragment depicts the agony of suddenly changed circumstances for the ancient Titans. The poem commences with a series of vivid images supported by stoic blank verse, the grounds for its popularity since its first publication in Keats's best-selling volume *Lamia, Isabella, The Eve of St. Agnes, and Other Poems* (1820). The "other poems" include all of Keats's odes (save "Indolence") and the unfinished *Hyperion*. The concentrated brilliance of his last two years of writing demonstrates his open experimentation with epistemology in poetry concerning nature. This ecological reading of *Hyperion* claims that the fragment is an early vision of the chaotic patterns that recent science has theorized as endemic to evolutionary history, most notably in the theory of evolution by punctuated equilibrium.[1] Oceanus's evolutionary philosophy is the teleological doctrine that "first in beauty must be first in might" (*Hyperion* 2.229), and critics have linked the narrative to the pattern of political revolution so pertinent to the Napoleonic era and its end in 1815 (Bewell, "Political Implications"; Levinson, *Keats's Life of Allegory* 208).

---

1. In 1972, Stephen Jay Gould and Niles Eldredge introduced the phrase "punctuated equilibrium" to scientific circles. Using the evidence of geology, the theory of punctuated equilibrium counters that of evolutionary gradualism in its claim that gaps in the fossil record show the true pattern of evolution as sporadic, with long periods of stasis, or equilibrium, in static environments interrupted by sudden bursts of evolution in response to changes in environment.

Ecological degradation is an equally compelling cause for the Titans' downfall that has little relation to the politics of Keats's time but considerable significance in the context of Keats's living environments. Keats was haunted by the prognosis of tuberculosis in his last years. The tubercular condition elicited medical representations of opposable environments: damp, chilly, foggy England versus dry, warm, sunny Italy; London hospitals crowded with convalescents versus open, airy escapes like Hampstead or Winchester; a cold rainy ride on a carriage roof versus the deep corporeal pleasure of a lingering autumn; an unlucky stroke or a string of good fortune. The quixotic character of nature in *Hyperion* embodies a modern understanding of atelic evolution riddled with contingency. Keats's *Hyperion* is a modern poem because the narrative of Providence and purpose quakes under his piled-up personal experience with bad-luck environments. Ecological mischance is a perversely formative force in the poem. Parallels abound between Keats's portrayal of the fallen Titanic world and Buffon's catastrophist primordial landscape. Buffon writes of the demise of gigantic ancient species outlined in the fossil remains: "The organic remains of land animals dispersed through this diluvial gravel must, with the greatest probability, be referred to the races extinguished by the great convulsion which formed that gravel; many of them are of species still inhabiting the countries where they are thus found, some of the species now inhabiting only other climates; and some few, of species and genera now entirely unknown" (32). Buffon presents his view of extinctions: "The great catastrophes which have produced revolutions in the basins of the sea, were preceded, accompanied, and followed, by changes in the nature of the fluid, and of the substances which it held in solution; and when the surface of the seas came to be divided by islands and projecting ridges different changes took place in every separate basin. [. . .] These interruptions and retreats of the sea have neither been slow nor gradual; most of the catastrophes which have occasioned them have been sudden" (35).

This natural history of sudden disaster is quite different from Werner's Neptunism or Hutton and Lyell's uniformitarianism. Buffon and his countryman Cuvier believed strongly in catastrophism as demonstrated by the fossils in the Paris Basin, and species extinction followed the logic of "great convulsion." Unlike Cuvier, Buffon believed in the mutability of species and integrated evolutionary speculation into his catastrophism, though the twenty-volume *Natural History* vacillates on these controversial ideas. By reading the most recent theories of how the natural world changes over time, Keats was able to translate contemporary science into a poetic myth that dramatized extinction and evolution, integrate his experience of polluted and miasmic environments into organism stress, and test out the tensile strength of his metaphysical struggle for existence in the "vale of Soul-making." *Hyperion* immediately delves "Deep in the shady sadness of a vale" (line 1). The dethroned and desolate Saturn is the first fossil the poem uncovers. The opening scene between Saturn and Thea is set in a despoiled landscape in which the silence following catastrophe reigns, where a "voiceless" and "deadened" stream passes fallen leaves and seeds.

> No stir of air was there,
> Not so much life as on a summer's day
> Robs not one light seed from the feather'd grass,
> But where the dead leaf fell, there did it rest.
> (1.7–10)

The light seed reappears from Keats's letter's as a seed fallen on the "wide arable land of events," a metaphor for Apollo, and the dead leaf is the fallen Saturn himself, soon to be stirred by Thea's plea to rage against the dying of the light.

In classic mythology, after losing to the Olympians in battle, the Titans were banished to Tartarus, a place of disrepute and degradation spawned from primordial chaos. Keats's poem is a narrative of interregnum describing the upheaval between two stable states—a mythological paradigm shift. Keats's scenes of action depict not the epic battles between two world orders, the new and the old, but instead the desperate, passive self-questioning that both Titans and Olympians face as a consequence of the new reality. As a work of epic narrative, *Hyperion* suffers from a dearth of action scenes because this poem is not about evolutionary competition but about luck, good and bad. The environment metes out fortune.

Rather than utterly differentiate the two classes of gods, Keats is careful to draw parallel scenes between the Titans, Saturn and Hyperion, and the new Olympian sun god, Apollo. This parallelism implies a genealogical relationship where power is passed down along kinship or evolutionary lines instead of a political vision influenced by the French Revolution (Little 140). Perhaps the most important of these parallels appears in line 103 in books I and III of the epic, when Keats's inverse heroes, Saturn and Apollo, seek to understand power in nature. The fallen Saturn asks in book I,

> Who had power
> To make me desolate? Whence came the strength?
> How was it nurtur'd to such bursting forth,
> While Fate seem'd strangled in my nervous grasp?
> (1.102–5)

The rising Apollo asks in book III,

> Where is power?
> Whose hand, whose essence, what divinity
> Makes this alarum in the elements,
> While I here idle listen on the shores
> In fearless yet in aching ignorance?
> (3.103–7)

The locus of power lies in controlling the chaos of the elements; when Saturn's family has suddenly lost this power, it turns out to have been a temporary gift of circumstance. Apollo receives power by having the disparate elements of history driven into his brain by Mnemosyne, the goddess of memory. Further contrastive parallels help us locate contingent dynamics in evolutionary time. Saturn asks for raw elements of chaos to make new the world, and Apollo receives prewoven narratives of history. In his eagerness to remold his lost identity from a deist's understanding of how nature works, Saturn asks,

> But cannot I create?
> Cannot I form? Cannot I fashion forth
> Another world, another universe,
> To overbear and crumble this to nought?
> Where is another Chaos? Where?
> (1.141–45)

Mnemosyne fills the cistern of lucky Apollo's brain with historical knowledge:

> Knowledge enormous makes a God of me.
> Names, deeds, gray legends, dire events, rebellions,
> Majesties, sovran voices, agonies,

Creations and destroyings, all at once
Pour into the wide hollows of my brain,
And deify me [. . .]
(3.113–18)

Saturn wants a handful of old chaos to work into a new kingdom; Apollo lies passive as the knowledge of history preceding him turns him into a god. The chaos of elemental creation that Saturn desires for his workshop is effectively translated into the chaos of contingent history in Apollo's brain, a history that requires acts of destruction before a new creation can be organized out of the elements. He is the only one, it seems, who will survive the "destroyings." Apollo leaves the degraded Titanic world when Mnemosyne promotes him as founder of the new realm, and it seems that the epic Titanic suffering is just a prequel to a show starring Apollo. But the fragment *Hyperion* ends, abruptly, in a stellar explosion: "And lo! from all his / Celestial * * * * * * * * * *" (3.135–36). It remains ambiguous whether the new god Apollo is promoted to a higher state or blown into bits of asterisk.

This is one of the more celebrated endings in a literary period known for fragmented narratives. Like a lexical supernova, the ending of *Hyperion* is the ultimate contingency. It reasserts chaos from the order of Apollo's body, demonstrating the final triumph of entropy. If this work is a coherent fragment, as Marjorie Levinson has claimed, it is because it ends with an explosion that burns away the dank despair of the Titans in the opening scenes (*Romantic Fragment Poem* 172–73). It is a reversal of cosmic history, starting with the dystopia of a destroyed landscape, ending with the big bang. This atomic dissolution is even more nihilistic than Saturn's plea to "overbear and crumble this [world] to nought." Keats's poem begins in a traditional mode of teleological evolution, with the outworn old realm falling to the innovative new power, but it smashes that coherent myth into an antilinguistic chaos of asterisks. Knowledge of history fails to make either class of gods more powerful or more effectual in their environments. Fate is merely what Keats called the "van of circumstance": radical contingency.

With so little agency afforded to the characters, the locus of power must lie elsewhere in this degraded world of the end-Titan era. In his study of epidemiological fear in the Romantic period, Alan Bewell mentions the ecological poisoning that sickens the Titans' environment. Bewell relates the Titans' sickness to early nineteenth-century visions of America that deeply interested Keats, who had lost one brother to tuberculosis and the other to the hazardous young United States (*Romanticism* 169). But human disease and miasma theory are concerns distinct from environmental toxicology. It is more evident that the Titans are poisoned by elements of pollution in their landscape, not by a virulent disease. Their physical suffering from toxic body load is complicated by the poisonous discourse of the desperate gods, who seek a source of power to blame for their suffering. In a vigil, Hyperion burns ceremonial incense, and "his ample palate took / Savour of poisonous brass and metal sick" (1.188–89). Incense is used in rituals of purification, meditation, and medicine. It is always composed of aromatic plant materials and essential oils, never of metals. This conflation of ritual organic burning and the combustion of alloy metals—compounds made possible by technology and industrial demands—is the first of a series of substitutions in which the Titans cultivate an alembic or ablution through traditional ceremony and instead find themselves sickened by the practice.

Heavy metal pollution is generated in the industrial processes of smelting and purifying, for example, mixing copper with zinc to manufacture brass. What the Titans were burning has relevance to modern industrial society, in Keats's era and the present. With the rapid increase of chemical and metal manufacturing driven and consumed by the British industrial economy, illnesses from exposure in the factory and in polluted environments became common. Legislation such as the Factory Acts of the mid-nineteenth century protected workers' health by limiting working hours per day, but did nothing to limit toxic exposure. Heavy metal pollution is particularly dangerous because heavy metal waste does not decay like organic pollution, and it is bioaccumulative: it enters the bodies of plants and animals and lies dormant. Metals are insidiously magnified by ingestion through the layers of the food chain, as we know from mercury content in large fishes such as tuna. Heavy metal pollution is often the culprit in industrial waste sites, where chronic or catastrophic leakage of by-products from metal smelting can contaminate large areas. These problems would start to receive widespread public attention only with the work of Rachel Carson in the second half of the twentieth century.

When Hyperion's "ample palate" tastes the "metal sick," Keats seems to be indicating the illness caused by acute exposure to toxic elements from the manufacture of alkali metals for industrial uses, including faintness, vomiting, and pulmonary disorders (Dingle 534). The process for producing alkali, invented by Nicholas LeBlanc in the late eighteenth century, emitted hydrochloric acid directly into the atmosphere surrounding the manufactory, sickening everything living in the vicinity, animal and vegetable. By 1862, alkali metal manufacture was called "the monster nuisance of all [industrial-era pollution]" (Dingle 529). Copper and zinc toxicity from the by-products of brass manufacture, fugitive dust, metal oxides, sodium fluoride, and formaldehyde can cause vomiting, irritation of skin and mucous membranes, lethargy, cancer, and mental derangement. Keats's atavistic poem on the ancient gods actually describes modern pollution sicknesses caused by the demands of industry for the raw materials of glass, textile, and soap production. The British in Keats's time were inundated with "metal sick."

The poisoned Hyperion rants against his fallen comrades, whom he calls "lank-eared Phantoms of black-weeded pools" (1.230). Hyperion sees them as degenerate—dirty and degraded sloths—rather than as sovereign gods. From the black pools where no birds sing, "A mist arose, as from a scummy marsh" (1.258). The miasmic mist of unhealthy vapors is the active agent in this dismal scene. Keats reuses a favorite and unusual adjective, "scummy," with a notable update from his usage in "Endymion": "To breathe away as 'twere all scummy slime / From off a crystal pool" (3.329–30). In the earlier poem, the "scummy slime" is a simile for burdened old age and the trance of years elapsed since youthful joy; it is breathed away by the arrival of Endymion. It is purely metaphorical scum, like a nightmare once one has awakened. In *Hyperion*, the scum has become disgustingly material. The "black pool," the "scummy marsh," and the unhealthy mist are the dystopian features of this new landscape, and the kingdom on the plain of Thessaly has devolved into a toxic environment.

As incense has turned to contaminant, crystal pools of Keats's earlier imagination have blackened, eutrophied, died. Pleasurable ritual has become a tormenting rehearsal of thanatos, the death drive, where life hastens toward dissolution. Environmental degradation becomes a nexus as the epidemiological circle of pollution closes in fatal sickness around the Titans. In a state of more than existential pain, the Titans' hearts are "Heaving in pain, and horribly convuls'd / With sanguine feverous boiling gurge of pulse" (2.27–28). These packed iambs force blood, unnatural heat, and constricted breath through their toxified bodies and seem to hasten death through the crisis of survival. It is a poignantly physiological description of bodily panic, where the very heat generated to kill off infection only weakens and attenuates the defended life—the destructive side of Keats's "ellectric fire." Saturn finds "a mortal oil upon his head, / A disanointing poison" (2.97–98), and the Titan dethroned in the opening scene becomes the sacrifice on the altar of circumstance. Apollo was to become god of medicine; he is an unlikely candidate for sickening the old gods. Keats inverts the Psalm of David: "Thou preparest a table before me in the presence of mine enemies; thou anointest my head with oil; my cup runneth over." Instead of the saints, kings, and priests, whose heads were anointed with olive oil to show their devotion and God's power to restore health, in Tartarus the oil is mortal, disanointing, poisonous, and delivered on the rod of "fate" rather than to chosen ones by a benevolent God. There could not be a more thorough set of reversals countering church teachings: Greek gods depose Christian monotheism; dystopian Tartarus washes away the delusion of Edenic utopia; a poisonous chaotic world shreds Providence.

Perhaps the Titans fouled their own nest, and the reader enters the scene when it is already a sickened vale, a poignantly modern industrial condition. The evidence of toxic dystopia in the "poisonous brass," the "black-weeded pools," the "scummy marsh," and "disanointing poison" shows how far Keats had driven his mind from the mellifluous scenes of "Endymion," a disavowal of that earlier mode of pretty versification that was so savagely derided by the critics. The dys-location in *Hyperion* intensifies the discombobulating, fragmented, antiteleological narrative that Keats sketched out in his contemporaneous letters. The Titans' untimely senescence is due to ecological pollution and toxic bioaccumulation, not a progressive vision of one species outcompeted by a superior species. Denise Gigante has theorized from these passages in *Hyperion* that Keats was developing an "allegory of taste" in which the Titans, as tragic heroes, are predisposed to feel disgust rather than pleasure (149). The visceral senses of smell and taste provide Keats, lover of poetic "gusto," with an intensive, synaesthetic image set for depicting epic agony in a revolting setting. The Titans' environment has degraded into a toxic habitat that infects their consciences and bodies in the new condition of mortality.

The concept of pollution was first used in English in the fourteenth century. Since then it has subtly evolved from its earliest denotation as spiritual or ceremonial desecration, as when the conscience is polluted by an evil act or the ears are polluted by blasphemy. Pollution came to be linked to physical contamination, and by the eighteenth century it had the definition we consider most relevant to modern concerns: to pollute the natural environment with the effluvia of industry: gas, liquid, and solid waste. By the nineteenth

century, especially in Keats's London, known as "The Big Smoke," water and
air pollution were quotidian observations. Keats's battle with tuberculosis,
previewed by the deaths of his mother and brother, forced him to develop an
industrial-era concern that spiritual and physiological contamination are
related. Especially considering his recent trauma with the death of his brother
and the miasmic theories of disease transmission circulated in his time, Keats
would have had ample reason to be apprehensive about the dangers of industrial damp environments and the people who fall sick within them.

Exacerbating the physical infirmities of the vulnerable race of Titans in
Hyperion are the environmental cataclysms that overtake their civilization.
Natural disasters like earthquakes, meteors, and tsunamis are distinct from
plagues and pollution, which are often wrought by the habits of the inhabitants. The Titans seem to be suffering from a coherent catastrophe involving
both habit and contingency. Keats consistently uses natural apocalyptic parallels to capture the condition of the Titans, which ingrains our sense that
environmental mischance is not just the metaphorical vehicle that drives
the plot forward, but also the conceptual foundation around which Keats
organizes evolutionary change:

> Blazing Hyperion on his orbed fire
> Still sat, still snuff'd the incense, teeming up
> From man to the sun's God; yet unsecure:
> For as among us mortals omens drear
> Fright and perplex, so also shuddered he—
> [.....................................]
> [. . .] horrors, portion'd to a giant nerve,
> Oft made Hyperion ache [. . .]
> [.....................................]
> His winged minions in close clusters stood,
> Amaz'd and full of fear; like anxious men
> Who on wide plains gather in panting troops,
> When earthquakes jar their battlements and towers.
> (1.166–70, 175–76, 197–200)

The blindness of dire apprehension, fed on toxic metallic fumes, is horrific
to a class of gods who have no history to consult beyond the stability of
their own long reign. In vain, the Titans seek deeper historical knowledge
to understand their place in the order of things (order is assumed), using
two distinct methods: Saturn's "old spirit-leaved book" and Oceanus's evo-

lutionary principle of beauty (2.133). Saturn's tome, a variant on natural theology, provides no perspective; he finds no reason why the Titans "should
be thus [fallen]" (2.149):

> Not in the legends of the first of days,
> Studied from that old spirit-leaved book
> Which starry Uranus with finger bright
> Sav'd from the shores of darkness [. . .]
> [. . . . .]
>   the which book ye know I ever kept
> For my firm-based footstool:—Ah, infirm!
> Not there, nor in sign, symbol, or portent
> Of element, earth, water, air, and fire,—
> At war, at peace, or inter-quarreling
> [. . . . .]
>   not in that strife,
> Wherefrom I take strange lore, and read it deep,
> Can I find reason why ye should be thus:
> No, no-where can unriddle, though I search,
> And pore on Nature's universal scroll
> Even to swooning, why ye, Divinities,
> The first-born of all shap'd and palpable Gods,
> Should cower beneath what, in comparison,
> Is untremendous might.
> (2.132–35, 137–41, 147–55)

Nothing in the recorded natural history of the Titans prepares them for
extinction, because their existence is relative to nothing else, they are "first-
born." They have no fossil record to consult. Their environment has never
before shown this degraded condition that seems to be in collusion with
Apollo to overthrow their reign. Saturn is destined to become that first fossil, as he is described embedded in the opening scene: "Forest on forest
hung above his head / Like cloud on cloud. No stir of air was there" (1.6–7).
Natural history will make an example of their sad case rather than providing them a context for evolutionary succession. Apollo subsequently inherits this history, and it annihilates him and the narrative itself.

Oceanus is a more patient student than Saturn. He looks to nature
rather than books, and his wise passiveness looks deeply into natural process. He sees an aesthetic improvement of forms over time. Oceanus begins,

> We fall by course of Nature's law, not force
> Of thunder, or of Jove. Great Saturn, thou
> Hast sifted well the atom-universe;
> But for this reason, that thou art the King,
> And only blind from sheer supremacy,
> One avenue was shaded from thine eyes,
> Through which I wandered to eternal truth.
> (2.181–87)

The light at the end of the tunnel is a backward perspective on Titanic origins: from "Chaos and parental Darkness came / Light," and the "atom-universe" falls into an increasing order and articulation in which each earlier stage clears the foundation and outlines the primitive form of the next (2.191–92). Oceanus's reassuring, if humbling, teleology spans the many forms of biological life:

> Say, doth the dull soil
> Quarrel with the proud forests it hath fed,
> And feedeth still, more comely than itself?
> Can it deny the chiefdom of green groves?
> [. . . . .]
> We are such forest-trees, and our fair boughs
> Have bred forth, not pale solitary doves,
> But eagles golden-feather'd, who do tower
> Above us in their beauty, and must reign
> In right thereof; for 'tis the eternal law
> That first in beauty should be first in might:
> [. . . . .]
> Receive the truth, and let it be your balm.
> (2.217–20, 224–29, 243)

The Titans are an impatient audience for Oceanus's evolutionary teleology. They are in no humor to swoon under the light of Apollonian beauty. Though its change toward ever greater beauty (or evolutionary "fitness") is provisionally comforting, muscular gods are not likely to fall without a fight, and no direct fight has ever been realized. The Titans spend many of their words fantasizing about a battle, but their number has fallen by the mediation of Nature, which cannot successfully be battled. Their environment falls to shambles around them, their health analogously decomposes, and the Titanic body seems destined to be recycled back into the "atom-universe." Oceanus's speech and his teleological principle of beauty are red herrings that distract from the contingencies that punctuate evolutionary history. Rather than end the poem with the realization of a higher order in the form of Apollo, Keats abandons it in the riot of the stars: "Celestial * * * * * * * * *" (3.136). It is time to reconsider this fragment as a completed turn in an evolutionary pattern that is punctuated and chancy. Keats remolds the mythology and geological catastrophism of his day to envision chaos at work in evolutionary ecology.

9

The intensive structure of the formal lyric poem is an interconnected system of words predisposed to support ecological dynamics in the poem's imagery. The interlaced rhymes, the default of iambic rhythm with some variation, and the repetition of the stanza footprint can be seen as prosody's "system," where the words and images are essentially more interreliant and synergetic than those of blank or free verse. The ode, as the most complex and ornate of the forms with which Keats tested himself, represents the lyrical high-water mark for his flood of inspiration in 1819. From spring to autumn of that year, Keats found his métier in this tangled, energetically self-organized form of ancient praise. Odes don't move like narratives do; they stay in a comfortable dwelling place and draw at the strings of the world around them. After the "Ode to Psyche," the first of his compositions in the spring of 1819, Keats devoted himself to a stable stanzaic structure in all the remaining odes, with ten-line stanzas, sonnetlike rhyming (a quatrain and a sestet), and pentameter lines. Exceptions prove the rule: the final "Ode to Autumn" has one additional line per stanza, and the "Ode to a Nightingale" has a single trimeter line. These gentle variations allow subtle distinctions among individual poems while reinforcing the evidence that the formal ode had a generative and balancing influence on Keats's verse.

New Critics praised the odes' formal perfection by finding the prosodic structure and gallery of images in synergy with the philosophical occasions for the work. This organic whole was seen as an object of aesthetic self-containment that kept the poems austere and separate from their context in Keats's society, politics, and view of history. These reverent close readings from critics like W. J. Bate, Helen Vendler, and Geoffrey Hartman have been complicated in the past twenty-five years by scholars of Romanticism interested in the historical and political valences at the edges of the odes. "To Autumn" can be historically contextualized by the Peterloo Massacre of 1819, which occurred four days after Keats's arrival in Winchester (McGann, *Beauty of Inflections* 58). In this light it is a revisionary response that contains and calms the bewildering political events of the late summer of 1819. "To Autumn" and the other odes can be read as poems about death (thanatopsis): the extinction of the urn's civilization, the tears of historical Ruth and her birdsong link to the emperor and clown and to Keats himself. With these poems, Keats is testing out the world as a vale of soul making by studying permanent symbolic objects as evidence of the transience of any one life, any one civilization. These are not poems breathing in their own rarefied, atemporal atmosphere. They allude to Keats's concerns about his own time in history, and the instinctual curiosity and precise description can be seen as empirical instincts (Chandler 404–30). These methods allow Keats to deduce principles by which natural history is organized.

"To Autumn" has already roused the attention of ecocritical scholar Jonathan Bate, as "a well-regulated ecosystem" and "an image of ecological wholeness which may grant to the attentive and receptive reader a sense of being-at-home-in-the-world" (106, 109). By studying natural systems acutely and in partial isolation, Keats is able to both create and describe a series of microcosms in verse. "Being at home in the world" is the trope of dwelling; what I see in these poems is an empirical vision of organic interconnection foreshadowing systems ecology. Seen as conceptual models, microcosms are even more revealing of ecological epistemology than Bate gives the poems credit for. Keats repeats the microcosm experiment six times in a single year, keeping relative controls over the prosody—only Psyche varies greatly in formal structure—and varying by subject, season, and personal mood. If one definition of British Romanticism is the literature of resistance or refiguration of the hegemonies of Christian doctrine and Enlightenment empiricism, we find these re-visions well established in Keats's work. His theory of secular salvation in the "vale of soul making" is his spiritual evolutionary hypothesis, and his theory of nonreductive empiricism is his holistic ecological hypothesis. This is Keats's delicate empiricism: the work of a chameleon poet who studies so stridently that he annihilates himself and grows into the color of his subject.

Since the chameleon poet was ready to blend into his surroundings and subjects to become one with them, passages in these poems frequently erase the speaker altogether, permitting a vision of the relations among nature's many selves. Keats's intense empathy with his subjects is known to be one reason why he found medical surgery of his day, performed without real anesthesia, to be intolerable in spite of his high aptitude for the work. The poet-scientist Goethe, one of continental Romanticism's great figures, theorized an empathic, noninvasive science he called "delicate empiricism, which makes itself utterly identical with the object, thereby becoming true theory. But this enhancement of our mental powers belongs to a highly evolved age" (Goethe 307). Keats's close attention to the forms and developments in nature could be seen as delicate empiricism for poetic ends, and some of his odes, particularly "Autumn" and "Nightingale," amount to lyrical natural histories. The words become an *oikos*, a dwelling place. The identity of the dweller is unimportant; simply being there is the occasion for the ode. Readers of Keats are easily able to squeeze themselves into his "I," and his frequent use of dream and drugged states of consciousness loosens identity from I to we, the we experiencing the poem. We're not certain we can trust what we see, smell, hear, feel, and taste, but the impression of dwelling in these aesthetic little worlds is so strong that truth may as well be the same thing as beauty.

These dwelling places embed the subject deeply within a natural habitat. Psyche and Cupid are "couched side by side / In deepest grass, beneath the whisp'ring roof / Of leaves and trembled blossoms, where there ran / A brooklet, scarce espied: / 'Mid hush'd, cool-rooted flowers, fragrant-eyed" (9–13); the lover of the nightingale sits "in embalmed darkness" with "The grass, the thicket, and the fruit-tree wild; / White hawthorn, and the pastoral eglantine; / Fast fading violets cover'd up in leaves; / And mid-May's eldest child" (43–48); the Grecian urn depicts "marble men and maidens overwrought, / With forest branches and the trodden weed" (42–43); the melancholy escapist is told to "glut thy sorrow on a morning rose, / Or on the rainbow of the salt sand-wave, / Or on the wealth of globed peonies" ("Ode on Melancholy" 15–17); the indolent soul becomes "a lawn besprinkled o'er / With flowers, and stirring shades, and baffled beams" ("Ode on Indolence" 43–44); and, in perhaps the best-measured natural system of the collection, the autumn afternoon holds the hum where

> the small gnats mourn
> Among the river sallows, borne aloft
> Or sinking as the light wind lives or dies;
> And full-grown lambs loud bleat from hilly bourn;
> Hedge-crickets sing; and now with treble soft
> The red-breast whistles from the garden croft;
> And gathering swallows twitter in the skies.
> ("To Autumn" 3:27–33)

Finding sensuous spaces in Keats is an easy task; what an ecological ode shows is that these spaces are not merely beauty for beauty's sake, or the escapist aesthetic ideals of adolescence, but close studies of exchange and scale in nature. The odes are microcosm experiments of the mind, meted out by the line and the rhyme.

This thesis is defensible partly because Keats's odes are more complex in their views of nature than the simple pastoral, sylvan, and great-house eclogues of his literary predecessors. Confusion, dissolution, and mortality play indispensable roles in Keats's visions of nature, and even the most vapid-seeming of celebratory lines ("More happy love! more happy, happy love!" from "Ode on a Grecian Urn") can be drawn into more penetrating light with an ironic reading (the insistent repetition of "happy" reveals how far from happy the speaker is, and the plosives "happy," like the word "pretty," dissever those words' meanings). Though joy and sadness have complex relations in these poems, Keats has no desire to engage in reductionism, which he calls "consequitive reasoning," an epistemology of the "sciential brain" that seeks, as he says in "Lamia," "To unperplex bliss from its neighbor pain; / Define their pettish limits, and estrange / Their points of contact" (1.191–94). His odes hold the wholeness of existence in nature as a dear value of the soulful individual, whose atoms of intelligence have been schooled by experience in "elemental space." Nowhere in his works does he seem closer to saving himself from oblivion in death, the great fear he had engraved on his tomb, than when his final ode closes with an atomic pattern of organization, where "gathering swallows twitter in the skies." With these words, the naturalistic impressions of a single man on an afternoon's walk in Winchester in the autumn of 1819 are embedded in the consciousness of Keats's inheritors. Not only students of literature ponder swallows: biologists and mathematicians have attempted to model the kaleidoscopic flocking patterns of swallows. Theories abound on how these patterns relate to survival by avoiding predators, achieving socialization, and enhancing seasonal migration. Keats's final ode image previews ecological perspectives on animal behavior within a time-space context.

The odes succeed because they lend us an intense vision of how nature can be perceived without reducing our experience to its literal components or overwhelming our minds with diluted universalities. Each ode finds its proper scope, then stays there to ponder a while. Part of this all-important scope, or circumscription, is assignable to form: ten- or eleven-line stanzas with sonnetlike rhyming, with the whole between thirty and eighty lines. But the images superadded to the formal system of prosody are also sheltered, veiled (vale-d), drawn within semitheoretical boundaries. The odes are poetic ecosystems, turned and gently manipulated by the voltas of mood and inspiration. Ecosystem science of the twentieth century would negotiate its terms with remarkably similar methods. Recall Arthur Tansley's theory of circumscription in ecology, which claims that ecosystems

> are of the most various kinds and sizes. They form one category of the multitudinous physical systems of the universe, which range from the universe as a whole down to the atom. The whole method of science [. . .] is to isolate systems mentally for the purposes of study, so that the series of *isolates* we make become the actual objects of our study, whether the isolate be a solar system, a planet, a climatic region, a plant or animal community, an individual organism, an organic molecule or an atom. Actually the systems we isolate mentally are not only included as parts of larger ones, but they also overlap, interlock and interact with one another. The isolation is partly artificial, but it is the only possible way in which we can proceed. (299–300)

The boundaries of ecosystem ecology are half-imagined for the sake of coherent study, and they half-exist as distinct substructures of organization in nature. Imagining their existence is a necessary condition for theorizing how they might work: how inclusive to be in the model (including the inorganic components of the system as well as the organic), how to measure the impalpable entities (like energy exchange), and how to account for unpredictable events set in time (droughts, storms, human impacts, climate shifts). The ecologist dwells in a liminal terrain between theory and material entity.

Keats's isolated small worlds are borrowed temporarily from larger places. He clears semitheoretical contemplative space away from, for example, "busy common-sense" ("Indolence" 40); he diverts from the path of emotional oblivion to stay in the experiential world with his plea "go not to Lethe" ("Melancholy" 1); he often sets himself up for loss in the end, when the theoretical circumscription melts back into the wide world outside of his ode system, as it does at the end of "Nightingale":

> Forlorn! the very word is like a bell
> To toll me back from thee to my sole self!
> [. . . . .]
> [. . .] adieu! thy plaintive anthem fades
> Past the near meadows, over the still stream,
> Up the hill-side; and now 'tis buried deep
> In the next valley-glades
> (71–72, 75–78)

The next valley over is another enclosed terrain, and the poet has no choice but to give freely what he never truly possessed. His surroundings have changed and the vision is gone, but thankfully the ode holds the moment in energetic stasis, mimicking an ecosystem.

Ecological microcosms require a spare diversity, where each essential component is represented and plays a particular role in the cycling dynamics of the system. Plants store solar energy in carbohydrates, consume carbon dioxide, emit oxygen, and regulate moisture in the soil and atmosphere. Animals consume carbohydrates and oxygen, emit carbon dioxide and nitrogenous waste, pollinate flowers, and disperse seeds. Soils store nutrients broken down by the death of complex forms, serve as a substrate for the growth of life, and regulate atmospheric gasses and moisture along local gradients. The atmosphere circulates essential gasses, maintains climatic homeostasis in the long term, and supports weather conditions in the short term. The sun energizes the system by directly stimulating photosynthesis in plants and warming the atmosphere by solar radiation. The processes of exchange, where energy and matter morph and find temporary stasis in some body, are more difficult to describe, but in the metaphor of the microcosm as a body, these processes are the blood. Real ecosystem dynamics are fantastically more complex than these brief descriptions. Microcosms, as models, strip down that overwhelming complexity to reveal the basic contributions of each component to the integrity of the system. That is why microcosms are so essential to ecological empiricism.

In Keats's mind, a naturalist empiricism emerges from the detailing of ecological diversity. The following stanza from the "Ode to Psyche" will serve as a test case for synergetic diversity:

> Yes, I will be thy priest, and build a fane
>> In some untrodden region of my mind,
> Where branched thoughts, new grown with pleasant pain,
>> Instead of pines shall murmur in the wind:
> Far, far around shall those dark-cluster'd trees
>> Fledge the wild-ridged mountains steep by steep;
> And there by zephyrs, streams, and birds, and bees,
>> The moss-lain Dryads shall be lull'd to sleep;
> And in the midst of this wide quietness
>> A rosy sanctuary will I dress
> With the wreath'd trellis of a working brain,
>> With buds, and bells, and stars without a name,
> With all the gardener Fancy e'er could feign,
>> Who breeding flowers, will never breed the same
> (50–63)

The "Ode to Psyche" is itself an exercise in poetry that imports complex nature into the defined confines of the cranium. Keats seeds his mind within concentric layers of the garden, the sanctuary, the fane, the dark-clustered trees, and the wild-ridged mountains. The Aeolian trope of inspiration shows "mind" as a slant rhyme with "wind"; "pleasant pain" facilitates the active development of biodiversity before the pain is assigned to the "working brain" of creativity; these natural elements between uncharted "mind" and "working brain," in effect, colonize a new ground and cultivate a garden of pantheistic, and psychological, worship. The working brain breeds flowers that have no name and are continually in evolution, a notable update of static Eden, where Adam named the unchanging features of a pet garden.

The poem's consistent pentameter, which Keats would vary more in later odes, has the breath to support this rich excess of cognition, the "zephyrs, streams, and birds, and bees," the "buds, and bells, and stars without a name," and of course the evolving flowers. It is an active stanza; the limp passivity that often weighs down Keats's subjects is passed on to the Dryads, and the poet/gardener exudes creative energy and control over the system. Pain is only pleasant to Keats when it is productive; like the sensual image of "aching Pleasure nigh" (23) in the "Ode on Melancholy," this enviable state of striving *and* accomplishing (by "breeding flowers" and "burst[ing] Joy's grape" ["Melancholy" 28]) yields a psychosomatic climax. The union of conscious effort with progress aligns with Keats's admission that this was the first poem that required "even moderate pains." The cognitive ecology of the passage, which is its own Italian sonnet within the ode, demonstrates how a brain might be worked into a self-sustaining, heterogeneous ecosystem that celebrates all the virtues of the human mind, just as a natural microcosm embodies the general dynamics of nature. A building is a place in which to dwell, and Keats's imagined temples and fanes clear space for worship and study set apart from the whole of nature. Keats wrote the earliest "Ode to Psyche" in a "more peaceable and healthy spirit" that would guide future writing efforts, and further delineates the odes period from the dark mood that drove the epic *Hyperion* (*Selected Letters* 294). Although this first ode to the "heathen Goddess" is one of his more uneven in structure and image, it sets the stage for this series of odes, each of which delineates a microcosm as the locus of attention.

The brain physiology becomes its own ecological-evolutionary system within the cranium, a distinct advance in the trope of the psychological microcosm. Keats continued to plow through permutations of the ode for the rest of that long summer. It is an insulated, prosperous niche in prosody and holds the seed of all natural potential yet unrealized, selectively breeding new species. We might name those flowers *Odus nightingalus*, *O. urnus*, *O. melancholus*, *O. indolencus*, and *O. autumnus*. Each poem finds a slightly different way to organize the same general prosodic scheme in genus, but the spots and stripes and colors of species contrast, highlighting the virtues of thematic variation. Natural systems are modeled in stanzaic formal structures where meter and rhyme exercise control and draw on affinities among the living components of the image.

The "Ode to a Nightingale" uses its first four stanzas, fully half of the long poem, to nudge away the sorrows and annoyances of life in the mainstream so as to settle in a self-contained space of nature and the bird, the promised "melodious plot" (8). The poet will not escape by using opiates or wine, agents of enervation and forgetting that work decidedly against perceiving any veritable reality. Instead, he pushes his unaltered brain toward the poetic station, where it "perplexes and retards" with the nightingale (34). This enables an alchemical insight that draws his visions from the clean celestial expanse down to the perplexing web of biological life on earth:

> tender is the night,
> And haply the Queen-Moon is on her throne,
>   Cluster'd around by all her starry Fays;
>   But here there is no light,
> Save what from heaven is with the breezes blown
>   Through verdurous glooms and winding mossy ways.
> (35–40)

With this ambiance set by the enclosed, welcoming gloom of a bower, the poet is positioned to model the natural world using his imagination. In terms of Tansley's "mental isolate," which draws a microcosmic construct between theory and reality, Keats articulates a haven of heterogeneous life borne on the boughs of ode-verse, and his musical bower becomes a resting place where he would be perfectly content to "become a sod" and forever dwell (60). He writes, in this complacent, dark stanza,

> I cannot see what flowers are at my feet,
>   Nor what soft incense hangs upon the boughs,
> But, in embalmed darkness, guess each sweet
>   Wherewith the seasonable month endows
> The grass, the thicket, and the fruit-tree wild;
>   White hawthorn, and the pastoral eglantine;
>   Fast fading violets cover'd up in leaves;
>   And mid-May's eldest child,
> The coming musk-rose, full of dewy wine,
>   The murmurous haunt of flies on summer eves.
> (41–50)

Time passes; this stanza marks the floral continuum from early spring all the way to full summer; and the fixed position of the speaker keeps his observations of the seasons coherent. Though the whole stanza is constrained by aporia framed in negative sensory terms ("cannot see" precedes "guess"), the speaker's familiarity with the "seasonable month" helps with the accuracy of his guesses about how nature is developing around him, from the grass and thicket to the rose and the flies. As in "Psyche," a list of entangled elements fills the sestet from "grass" to "summer eves," and "mid-May's eldest child," set alone in trimeter, sounds backward by rhyme to "the fruit-tree wild" and forward by consonance to the "coming musk-rose" and the "murmurous haunt." This wild eldest child, which is the musk-rose and an allusion to Keats himself, murmurs the infantile M and enters a transcendent state of meditation. It is a world the poet knows well enough to see without seeing it because the other senses are enough; he has turned the bower into a concept.

Keats's ecosystem holds its essential components: an observer (the poet), an occasion or subject (the nightingale), and its biotic and abiotic medium (the flowers, the leaves, the light, the sod). Without the nightingale, Keats would have no object of focus; it is the charismatic fauna that so often focuses conservation efforts on a particular place. The reaches of the bird's song are the edges of the poem. Although this suburban plum-tree bower is not endangered in the modern sense, it is assuredly a little haven set amid the slightly larger haven of Hampstead, which lies within sprawling industrial London. Self-sufficient and yet vulnerable, the nightingale bower occupies real and symbolic worlds simultaneously. The specifics, the flora and fauna of this space and their particular interactions, are accessible by scale and enable extrapolation onto larger places like the near meadows and the next valley glades. The speaker wishes to "become a sod"—to be integrated with the earth as a strong Saxon noun while listening to the Latinate "requiem" of the bird's elaborate song. The nightingale is a fleeting visitor, and the poem's final seam is sewn with the bird's flight and the poet's wakening back to the quotidian world. The circle of song is broken, and the poet is forlorn—so forlorn that we carry a sense that the entire complex scene has fallen away with the loss of the bird. Even amid the new silence, though, the poet remembers and records his vision, which has become an archetype of the contemplative lyric. A plum-tree bower balances a complete little world through a fleeting eighty lines.

The final ode, "To Autumn," is known as the most perfect of the odes, the one that leaves the fewest open questions, the most balanced in stanza, movement, image, the most settled in its contentment with place. It is a fitting work to finish the clutch, a paused balance between life and death that celebrates a season turning back to darkness. It is a poem that prizes natural processes as indefinite rather than finite, and it impresses images of real, material nature more than the other odes, which circle around escapist ideals and abstractions. As an embedded, earthy, oozing lyric, it reveals itself as the product of the poet's experience over the course of a year of writing about small-scale nature.

Jonathan Bate's reading traces the string of good weather in relation to Keats's moods and his enthusiasm for the riparian walk out of Winchester (106–9). Finding respite from the perpetual chill of a tubercular patient, as well as delighting in the best weather since the Tambora volcano had caused the year without a summer in 1816, Keats delights in the easy overabundance of nature that draws humans into its open rhythms rather than sequestering them in man-made, fire-warmed, smoky inside spaces. One way this ripe overabundance is attained is by his addition of an eleventh line to each stanza; all three eleventh lines are images of excess fecundity ("summer has o'erbrimm'd their clammy cells" [11]; "Thou watchest the last oozings hours by hours" [22]; "gathering swallows twitter in the skies" [33]). Though the stanzas are highly self-contained, one slant rhyme threads through the whole work, joining "core" and "more" (in the first stanza) to "store" and "floor" (in the second) to "mourn" and "bourn" (in the final); this affinity of sound escorts the poem through three stages of production (more), inventory or possession (store), and depletion (mourn). These seasonal movements shadow the three-part mood of the ode, the triptych of midyear dissolving into late year, and nature's and the poet's resistance to this inevitable dissolution.

Human figures only half-exist in the poem: the addressee is autumn itself, which takes the various forms of "bosom-friend" (2) and conspirer with the sun, winnower, reaper, and gleaner; humans share the scene equally with bees, birds, gnats, lambs, crickets. The somnolent state that transposes imagination and reality is here exported to his figures of nature, and the observer remains austere, like the ideal scientist, acting only as observer behind the lines that conjoin at rhymes and part ways by stanzaic decree. The speaker asks important rhetorical questions ("Where are the songs of spring?" [23]), and his position remains that of a reader of a pleasant chapter from the book of nature because the setting has an infused instinct for call-and-response. The songs of spring are gone, but the music of autumn is a symphony: a wailful choir of gnats, loud-bleating lambs, singing crickets, whistling redbreasts, and twittering swallows. The atmosphere circulates from "mists" (1) to "barred clouds" (25), the fields from "sweet kernel" (8) to "stubble-plains" (26). The poet vividly observes these evolutions while preserving the static suspension of a fully

realized natural system, one that has been in the works since the chilly
greens of spring.

> Season of mists and mellow fruitfulness,
> 　　Close bosom-friend of the maturing sun;
> Conspiring with him how to load and bless
> 　　With fruit the vines that round the thatch-eaves run;
> To bend with apples the moss'd cottage-trees,
> 　　And fill all fruit with ripeness to the core;
> 　　　　To swell the gourd, and plump the hazel shells
> 　　With a sweet kernel; to set budding more,
> And still more, later flowers for the bees,
> Until they think warm days will never cease,
> 　　For summer has o'er-brimm'd their clammy cells.
>
> Who hath not seen thee oft amid thy store?
> 　　Sometimes whoever seeks abroad may find
> Thee sitting careless on a granary floor,
> 　　Thy hair soft-lifted by the winnowing wind;
> Or on a half-reap'd furrow sound asleep,
> 　　Drows'd with the fume of poppies, while thy hook
> 　　　　Spares the next swath and all its twined flowers:
> And sometimes like a gleaner thou dost keep
> 　　Steady thy laden head across a brook;
> 　　Or by a cider-press, with patient look,
> 　　　　Thou watchest the last oozings hours by hours.
>
> Where are the songs of Spring? Ay, where are they?
> 　　Think not of them, thou hast thy music too—
> While barred clouds bloom the soft-dying day,
> 　　And touch the stubble-plains with rosy hue;
> Then in a wailful choir the small gnats mourn
> 　　Among the river sallows, borne aloft
> 　　　　Or sinking as the light wind lives or dies;
> And full-grown lambs loud-bleat from hilly bourn;
> 　　Hedge-crickets sing; and now with treble soft
> 　　The red-breast whistles from a garden-croft;
> 　　　　And gathering swallows twitter in the skies.

(1–33)

In the first stanza, the weather conditions (mists and sun) find direct translation by rhyme and word proximity into the energy exchange that promotes plant growth (load, bless, bend, swell, set). The sun provides the vines in fruit with the energy to run; the mists and mellowness of the atmosphere provide the gourd and hazel shells with conditions to swell and plump. The cottage is wedged between moss and trees. The hazel shells and clammy cells, by rhyme and placement, fall in parallel as the seasonal development of plant and animals alike. As in the "Ode to a Nightingale," where the "murmurous haunt of flies" follows a menagerie of flowering plants, here the pollinators are bees drunk in the nectar of their natural work, still busily gathering honey for their overflowing hives and therefore still fusing the genes of squash vines, apple and hazel trees. Gifts are exchanged in a gratuitous excess among the weather, the plants, and the animals. The complex system is balanced in temporal suspense by the season's ongoing cooperation—a utopian vision of unusually warm weather in the gracious lingering autumn of 1819.

The second stanza personifies the season as a series of earth laborers whom Keats happily does not designate as dryads or mythological gods. The winnower is the wind; the reaper is winter delayed; the gleaner is a warbling stream. Exchanges occur between these abstract humanoid forces of nature and the plants they modify: wheat into grain, poppies into soporific, apples into cider. The products of culture—grain, drugs, cider—are so integrated into the landscape that human production and natural process blend into one. Where no reaper exists, the "half-reap'd furrow" is a natural disturbance, and the next swath remains an intact plot of grain and twined flowers. Without an active gleaner, the oozing apples lie uncollected beneath the tree. Keats subtracts out human agency to energize the natural system, so these phrases become ecological processes.

In the final stanza, climate and seasonal process are exhibited in the actions of animals, and the brown font predominates. It is the stanza of adieu, where "warm days will never cease" changes to "barred clouds bloom the soft-dying day." The season is, after all, mortal, and the animals are engaged in their species-specific choruses of goodbye by wailing, bleating, singing, whistling, and twittering. The ecological stage is a diverse landscape with plain, sallow, bourn, croft, and sky, each providing a niche for an animal type: insects in the sallow, ruminants on the bourn, birds resting in the croft and gathering in the sky. The forces of the setting sun and light wind both extend the feeling of time suspended and forecast the turn toward winter. The autumn in microcosm is robust enough to sustain the scene independently of human energies and demands. Although the poem is the product of a walk on a single afternoon, it possesses the wisdom of a full cycle of life, where another year has been fostered by birth, growth, production, and reproduction. Seasonally, it joins with the springtime "Ode to Psyche," in which Keats vowed to build a fane to his brain and breed flowers of insight, as he has done by September.

Because he does not push autumn over the equinoctial edge, Keats's poem stays living in this state of dynamic suspense, "gathering" its energies like a cloud of swallows in the "skies." The final word recalls its rhyme-pair from where "the light wind lives or dies," and the skies, like the ode itself, cradle life and death in the same cornucopia. The scene that is suspended in the last line draws a trophic link between swallows, which eat the gnats, which eat the fruit oozing in the earlier stanzas. Swallows bring the global implications to this small scene: they migrate from northern Europe to southern Europe and Africa in yearly patterns. Their gathering in the skies is a harbinger of the population's exodus to warmer climates for the coming winter season. But the rupture of this ecosystemic circle is forever suspended, and the scene, auditorily sustained by rhyme and meter, is ecologically sustained by the cosmic balance of Keats's images, flora with fauna, light with darkness, life with death. The swallows are forever going, but they are never gone; the microcosm holds time and space perpetually.

The odes clearly oppose the ruptured natural history of *Hyperion* with self-enclosed, self-sustaining modeling schemes of the natural world. Keats pushes the narrative and lyric forms into intriguing ecological case studies. The strength of the odes is borne on their scope, their organization, and their use of prosody to support systems thinking. Keats's poems preview questions that ecological science would broach in the next hundred years. If ecosystems are partly conceptual, how can the union of concept with material construct enable greater insight than reductionism? Is narrative or modeling a more powerful tool in ecological empiricism? When dealing with complex, irreducible systems that change over time, what is the proper perspective or level of analysis? Who is the perceiver? Is our perception of natural beauty—stability, symmetry, diversity, adaptation—the truth of nature, or do we just imagine balance to forestall fears of natural disorder? Is truth not beauty but chaos?

The mass industrialization of the nineteenth century was perhaps the greatest revolution in human history. Our lives bear little resemblance to those lived only two hundred years ago, or one one-thousandth of the time *Homo sapiens* have dwelled on earth. This revolution has caused acute environmental degradation across the range of global ecosystems. Mostly, the transformation of wild landscapes into places that serve human needs has gone unchecked, and the cultures whose dwelling is ecophilosophically gentler have been on the decline. Nineteenth-century writers provided the early impetus for understanding ecological problems, and alleviating some of them, by developing conceptual techniques for investigating nature. Romantic and Victorian perspectives on ecology were often strained between the aesthetic appreciation of natural beauty and the scientific investigation into how natural systems work. This study has sought to resolve these contrasting perspectives by proposing that two essential tropes provided both the aesthetic and the empirical grounds for study. Extending this relationship is the complementary role of microcosm as model and chaos as a narrative practice. Although the tropes have contrasting epistemological forms, they repeatedly rely on each other for meaning. Chaos is demonstrated by running models; the microcosm's state vacillates between stability and chaotic change.

I chose to focus on British literature because England was the first industrialized country, so it experienced some of the first industrial-era shifts in climate and landscape. Another supporting factor is that the literature and science of Great Britain represent the vanguard of Western imperial knowledge in the nineteenth century. The British invested heavily in institutes of learning and exploration, so they had a uniquely global perspective on nature. I have analyzed innovative novels that envision contingency in nature and drive their plots using the radical notion of evolutionary punctuations, rather than Darwin's coherent gradualism, in the narratives of nature over time. I have identified the conceit of the microcosm as a way to delimit and organize nature, and shown how it was a particularly powerful trope in formal lyric poetry owing to the interdependent system of prosody. These two perspectives on chaotic narrative and microcosmic lyricism suggest that the tropes were useful tools for analyzing nature's behavior, and I've noted affinities between these works and methods in modern ecological science.

Studies of literature are often confined to the appreciation of fine writing that illuminates a bygone era of history or explores the perennial complexities of the human condition. The literature discussed in this book should not be limited to past time or a purely humanist approach. The point is not just that Romantic and Victorian writers achieved early empirical visions of nature that anticipated ecological methods in the centuries to come. There is more. The literary imagination at play in Keats, Wordsworth, Jefferies, Wells, and the others, and others I have not discussed in this book, shows how epistemological methods in literature can often achieve an impression of nature that is more visceral and inhabitable, and therefore more valuable, than the reductive methods of the ecological sciences. There is at once continuity between the tropes in literature and the sciences and clear distinctions between their definitions and uses. Where scientific modeling provides essential information about the behavior of systems under certain conditions, it tends toward simplification, reduction, and abstraction. Scientific modeling can actually enervate the natural system that is its inspiration.

Literary modeling does the opposite: it maintains complexity, holism, and in situ context, and it tends toward enhancing the vitality of the system it describes. Literary microcosms choreograph the discrete components of a natural system and leave space for the little wildernesses that lurk within the model. Keats's walk along the river Itchen near Winchester resulted in a model poem on a bucolic landscape in autumn that is a standard of its form, the ode, and also the skin in which that particular ecosystem continues to live to this day. New readers unwrap, examine, and play with the model every day. Literary narrative, especially in fiction, has the power to test out chaotic change in nature across a variety of scenarios and at different points in history. Where classic narratives are linear, coherent, and tend to resolve tensions in a satisfying conclusion, some nineteenth-century writers saw that natural degradation and disaster called for a different style of writing that focused on contingency and final indeterminacy. That more modern plot line is the only honest way we can tell stories about twenty-first-century nature. Both literary models and narratives come in a form that is accessible and appealing to a wide audience, and so hold the power of instruction and consensus building.

Because of their dedication to nature and their historical position on the nose of the oncoming Industrial Revolution, the Romantics and their Victorian-era compatriots are an exceptional group for ecoscience studies. However exceptional, they are far from exclusive. Numerous other literary eras would reward this kind of interdisciplinary ecoscience scholarship. Writers who grew up in the age of ecology, the early to mid-twentieth century, such as A. R. Ammons, Jacquetta Hawkes, Wallace Stegner, Gary Snyder, John McPhee, and Annie Dillard, were literate in the science of their time and used their work to deepen and complicate the theories of their scientific contemporaries. For example, Ammons's "Corson's Inlet" toys openly with chaos theory in its images and repetitions, and thus paints the model of the ever-morphing seashore in fractal patterns. In *A Land*, Jacquetta Hawkes uses her training in geology to envision a history of Britain in layers of landscapes and cultures worn under. England in 1949 was a perfect vantage point for reflecting on chaotic ecocultural change as the dust settled from World War II. The others I mention above use poetic narrative to attain fresh ecological perspectives on intimately known places. Each gives a unique contribution to literary empiricism.

Contemporary writers are the most obviously equipped to use literary methods as the epistemology of modern nature. Unable to ignore the influence of science, mass media, consumer habits, and the ecoethics of globalism, writers like Barbara Kingsolver, Margaret Atwood, Michael Pollan, and Bill McKibben have invigorated today's debates about environmental ethics because their writing is accessible and relevant to the lives of their readers. Books like Kingsolver's *Prodigal Summer* and Atwood's *Oryx and Crake* have often been the subject of ecocriticism articles and conference papers. Rarely do these critiques discuss how the books engage with evolution, contingency, and systems theory in material nature. In *Prodigal Summer*, how do Deanna's coyotes relate to Lusa's goats, and what effect does human stewardship of threatened species (coyotes and chestnuts) have on the evolutionary ecology of the mountain? What about Atwood's proposal in *Oryx and Crake* that the Internet and biotechnology are the vectors of climate change and pandemic, a punctuation point in evolution that clears the landscape for Crake's eugenics? Are biotech chimeras a realistic future that should be considered in community ecology? Like any high-quality science fiction, these ecological novels are predictive models that provide insights about the behavior of nature in response to human impacts.

Folk knowledge is often buried under hegemonic science, but as the voices of marginalized groups have found a wider audience with the spread of global literary study, the native ecological knowledge of those groups can enlighten and enrich the scientific study of global ecosystems. For example, Marilou Awaikta's *Selu: Seeking the Corn-Mother's Wisdom* records Cherokee traditions of knowledge and extends their philosophy of stewardship to today's environmental problems, and her *Abiding Appalachia: Where Mountain and Atom Meet* is a journey along the trail of the Cherokee spiritual leader Little Deer inspired by her youthful proximity to Oak Ridge, Tennessee, where in the 1940s the Manhattan Project advanced the destructive capabilities of nuclear science within one of the world's oldest mountain ranges. Awaikta's Native American knowledge combines an ethic of respect for nature and stewardship of ecological balance with awe for the apocalyptic forces that can be liberated by science. Alexis Wright's *Carpentaria* explores aboriginal Australian conflicts with white inhabitants of Queensland and multinational mining interests. Other native writers, particularly those inspired by colonial-era impacts on their home ground, have contributed valuable narratives of nature before, during, and after the arrival of profit-driven science.

What about all the writers who lived before the Industrial Revolution and before evolutionary and ecological science? Although theories of nature in, for example, medieval Britain were overwhelmingly static-state creationist, there are still valuable observations about deforestation, the spread of disease, the decline of overhunted game, and the vagaries of the annual harvest that can shed light on prescientific ecological perception. Recent research has uncovered observations in the Irish annals of the cold weather subsequent to volcanic eruptions (McGrath). The annals were kept by monastic scribes who recorded significant events in the culture and nature of Ireland between A.D. 431 and 1649. By comparing the annals' entries with ice core samples that preserve the chemical markers of eruptions, researchers have shown that the annals noted cold weather in the years following thirty-seven eruptions, including frozen lakes, crop failures, and human and livestock deaths. This research lends further support to ice core sampling as a way of profiling past climate and modeling the impacts of future eruptions and climate changes. It also shows how extreme weather events were associated with religious omens of apocalypse, perhaps the first wave of chaos ecology in human theories of nature. The annals and other ancient texts provide irreplaceable data that can be used to develop profiles of bygone ecosystems and the cultures that subsisted in them. If past is prologue, if the industrial era is indeed an energetic blip on the long line of ecohistory, we will benefit from recovering these preindustrial narratives and weaving them into complementary scientific research.

Tracing specific relations of influence between disciplines, countries, and centuries is an elusive goal. Instead, I hope to have demonstrated that both the humanities and the sciences have sought answers to how industrialism might change future conditions for our species and all the others. The insight pursued by environmental narratives and lyrics of the nineteenth century is essential to humanistic origins of ecological knowledge in the modern industrial world. By legitimating the power of creativity and imagination not only in literature but also in science, we dislodge a few bricks from the walls between specialized disciplines, and highlight the necessity of collective enterprise when it comes to facing off the looming environmental problems of today. The more we appreciate that the ecology of this century is more than just a scientific discipline, the closer we come to an intercultural investment in our global home.

Works Consulted

"Acid Rain." *The NASA Earth Observatory Glossary.* EOS Project Science Office, NASA Goddard Space Flight Center, n.d. Web.

Allard, James R. *Romanticism, Medicine, and the Poet's Body.* Aldershot: Ashgate, 2007. Print.

Allen, David E. *The Naturalist in Britain: A Social History.* London: A. Lane, 1976. Print.

Allen, Timothy F. H., A. J. Sellmer, and C. J. Wuennenberg. "The Loss of Narrative." *Ecological Paradigms Lost: Routes of Theory Change.* Ed. K. Cuddington and B. Beisner. London: Elsevier Academic, 2005. 333–70. Print.

Arata, Stephen. *Fictions of Loss in the Victorian Fin-de-Siècle.* New York: Cambridge UP, 1996. Print.

Argyros, Alex. *A Blessed Rage for Order: Deconstruction, Evolution, and Chaos.* Ann Arbor: U of Michigan P, 1991. Print.

Arnold, Matthew. *Poems.* Ed. M. Allott. Oxford: Oxford UP, 1995. Print.

Atkinson, A. D. "Keats and Kamchatka." *Notes and Queries* 196 (1951): 340–46. Print.

Atwood, Margaret. *Oryx and Crake: A Novel.* New York: Nan A. Talese, 2003. Print.

Bailey, G. H. "The Air of Large Towns." *Science* 13 Oct. 1893: 201–2. Print.

Barkan, Leonard. *Nature's Work of Art: The Human Body as Image of the World.* New Haven: Yale UP, 1975. Print.

Bate, Jonathan. *The Song of the Earth.* Cambridge: Harvard UP, 2000. Print.

Bate, Walter J. *John Keats.* Cambridge: Harvard Belknap, 1963. Print.

Beer, Gillian. *Darwin's Plots: Evolutionary Narrative in Darwin, George Eliot, and Nineteenth-Century Fiction.* Cambridge: Cambridge UP, 2000. Print.

Bellanca, Mary Ellen. *Daybooks of Discovery: Nature Diaries in Britain, 1770–1870.* Charlottesville: U of Virginia P, 2007. Print.

Bewell, Alan. "The Political Implications of Keats's Classicist Aesthetics." *Studies in Romanticism* 25 (Summer 1986): 220–29. Print.

———. *Romanticism and Colonial Disease.* Baltimore: Johns Hopkins UP, 1999. Print.

Beyers, Robert J., and Howard T. Odum. *Ecological Microcosms.* New York: Springer-Verlag, 1993. Print.

Boia, Lucien. *Weather in the Imagination.* London: Reaktion, 2005. Print.

Botkin, Daniel. *Discordant Harmonies: A New Ecology for the Twenty-First Century.* Oxford: Oxford UP, 1990. Print.

Bowler, Peter J. *Evolution: The History of an Idea.* Berkeley: U of California P, 2003. Print.

———. *The Norton History of the Environmental Sciences.* New York: W. W. Norton, 1993. Print.

Bowler, Peter J., and I. R. Morus. *Making Modern Science: A Historical Survey.* Chicago: U of Chicago P, 2005. Print.

Blake, William. "Preface to 'Milton: A Poem.'" *English Romantic Writers*. By David Perkins. 2nd ed. New York: Harcourt-Brace, 1995. 152–91. Print.

Brady, Ronald H. "The Idea of Nature: Rereading Goethe's Organics." *Goethe's Way of Science: A Phenomenology of Nature*. Ed. D. Seamon and A. Zajnoc. Albany: SUNY P, 1998. 83–114. Print.

Briffa, K. R., P. D. Jones, F. H. Schweingruber, and T. J. Osborn. "Influence of Volcanic Eruptions on Northern Hemisphere Summer Temperature Over 600 Years." *Nature* 393 (1998): 450–55. Print.

Buell, Frederick. *From Apocalypse to Way of Life: Environmental Crisis in the American Century*. New York: Routledge, 2003. Print.

Buffon, George-Louis Leclerc, Comte de. *Natural History, General and Particular*. Trans. William Smellie. 20 vols. London: T. Cadell and W. Davies, 1812. Print.

Burke, Edmund. Excerpt from *A Philosophical Enquiry into the Origin of Our Ideas of the Sublime and Beautiful*. *The Longman Anthology of British Literature*. Vol. 2A, *The Romantics and Their Contemporaries*. Ed. Susan Wolfson and Peter Manning. New York: Longman, 2003. 499–505. Print.

Busch, Akiko. "Romancing the Random." *Utne Reader* Jan.–Feb. 2011: 75. Print.

Campbell, Thomas. "The Last Man." *The Poetical Works of Thomas Campbell*. London: Edward Moxon, 1837. 104–7. Print.

Carpenter, Stephen R. "Microcosms Have Limited Relevance for Community and Ecosystem Ecology." *Ecology* 77.3 (1996): 677–80. Print.

Carroll, Joseph. "Human Nature and Literary Meaning: A Theoretical Model Illustrated with a Critique of *Pride and Prejudice*." *The Literary Animal: Evolution and the Nature of Narrative*. Ed. J. Gottschall and D. S. Wilson. Evanston: Northwestern UP, 2005. 76–106. Print.

Cervelli, Kenneth. *Dorothy Wordsworth's Ecology*. New York: Routledge, 2007. Print.

Chandler, James. *England in 1819: The Politics of Literary Culture and the Case of Romantic Historicism*. Chicago: U of Chicago P, 1998. Print.

Chesson, Peter. "Understanding the Role of Environmental Variation in Population and Community Dynamics." *Theoretical Population Biology* 64.3 (2003): 253–54. Print.

Chico, Tita. "Minute Particulars: Microscopy and Eighteenth-Century Narrative." *Mosaic* 39.2 (2006): 143–61. Print.

Christianson, Gale. *Greenhouse: The 200-Year Story of Global Warming*. New York: Walker, 1999. Print.

Clare, John. *Poems by John Clare*. Ed. Arthur Symons. London: Henry Frowde, 1908. Print.

Clements, Frederic. *Plant Succession and Indicators*. New York: H. W. Wilson, 1928. Print.

Coleridge, Samuel Taylor. *Coleridge's Table Talk*. Ed. John Potter Briscoe. London: Gay and Bird, 1899. Print.

——. *The Complete Poetical Works of Samuel Taylor Coleridge*. Ed. Ernest Hartley Coleridge. Oxford: Clarendon, 1912. Print.

——. "The Statesman's Manual (1816)." *English Romantic Writers*. By David Perkins. 2nd ed. New York: Harcourt-Brace, 1995. 618–19. Print.

Conte, Joseph M. *Design and Debris: A Chaotics of Postmodern American Fiction*. Tuscaloosa: U of Alabama P, 2002. Print.

Cooke, Brett, and Fredrick Turner, eds. *Biopoetics: Evolutionary Explorations in the Arts*. Lexington: ICUS, 1999. Print.

Cooper, Gregory J. *The Science of the Struggle for Existence*. Cambridge: Cambridge UP, 2003. Print.

Cox, Jeffrey. *Poetry and Politics in the Cockney School: Keats, Shelley, Hunt, and Their Circle*. Cambridge: Cambridge UP, 1998. Print.

Crabbe, George. "The Poor and Their Dwellings." *The Poetical Works of the Rev. George Crabbe: With His Letters and Journals, and His Life.* Vol. 3. London: John Murray, 1834. 283–98. Print.

Cracknell, Basil. *Outrageous Waves: Global Warming and Coastal Change in Britain Through Two Thousand Years.* Chichester: Phillimore, 2005. Print.

Cronon, William. "A Place for Stories: Nature, History, and Narrative." *Journal of American History* (Mar. 1992): 1347–76. Print.

———. "The Trouble with Wilderness; Or, Getting Back to the Wrong Nature." *Uncommon Ground: Rethinking the Human Place in Nature.* Ed. William Cronon. New York: W. W. Norton, 1996. 69–90. Print.

Csala-Gati, Katalin, and Janós Tóth. "The Socio-Biological and Human-Ecological Notions in *The Time Machine.*" *Wellsian: The Journal of the H. G. Wells Society* 26 (2003): 12–23. Print.

Dalmedico, Amy. "Models and Simulations in Climate Change: Historical, Epistemological, Anthropological, and Political Aspects." *Science Without Laws: Model Systems, Cases, Exemplary Narratives.* Ed. A. Creager, E. Lunbeck, and M. N. Wise. Durham: Duke UP, 2007. 125–56. Print.

Darwin, Charles. *The Formation of Vegetable Mould Through the Action of Worms.* Chicago: U of Chicago P, 1985. Print.

———. *On the Origin of Species.* New York: Oxford UP, 2008. Print.

———. *The Voyage of the Beagle.* Ed. Janet Browne and Michael Neve. London: Penguin, 1989. Print.

Darwin, Erasmus. *Zoonomia.* New York: AMS, 1974. Print.

Dawkins, Richard. *The Selfish Gene.* New York: Oxford UP, 1976. Print.

de Almeida, Hermione. *Romantic Medicine and John Keats.* New York: Oxford UP, 1991. Print.

de Quincey, Thomas. *Confessions of an English Opium-Eater.* New York: Penguin, 2003. Print.

Diamond, Jared. *Guns, Germs, and Steel: The Fates of Human Societies.* New York: W. W. Norton, 1997. Print.

Dickens, Charles. *Bleak House.* London: Bradbury & Evans, 1853. Print.

Dingle, A. E. "'The Monster Nuisance of All': Landowners, Alkali Manufacturers, and Air Pollution, 1828–64." *Economic History Review* ns 35.4 (1982): 529–48. Print.

Dodson, Stanley. *Introduction to Limnology.* New York: McGraw-Hill, 2005. Print.

Drake, J. A., G. R. Huxel, and C. L. Hewett. "Microcosms as Models for Generating and Testing Community Theory." *Ecology* 77.3 (1996): 670–77. Print.

Drury, William H., and Ian Nisbet. "Succession." *Journal of the Arnold Arboretum* 54 (1973): 331–68. Print.

Eldredge, Niles. *Life in the Balance: Humanity and the Biodiversity Crisis.* Princeton: Princeton UP, 1998. Print.

Eliot, George. *Middlemarch.* Oxford: Oxford World's Classics, 1998. Print.

———. *The Poems of George Eliot.* Ed. William Brighty Rands. New York: T. Y. Crowell, 1884. Print.

EPA (U.S. Environmental Protection Agency). "Chapter 3: Water Quality Criteria (40 CFR 131.11)." *Water Quality Handbook.* Washington, D.C.: U.S. Environmental Protection Agency, Mar. 2012. Web.

Ewing, Heather. *The Lost World of James Smithson.* New York: Bloomsbury, 2007. Print.

Forbes, Stephen. "The Lake as a Microcosm." *Bulletin of the Scientific Association* (1887): 77–87. Print.

Foucault, Michel. *The Order of Things: An Archaeology of the Human Sciences.* New York: Vintage, 1970. Print.

Fourier, Joseph. "General Remarks on the Temperature of the Terrestrial Globe and the Planetary Spaces." Trans. Ebenezer Burgess. *American Journal of Science and Arts* 32 (July 1837): 1–19. Print.

Fraistat, Neil. *The Poem and the Book: Interpreting Collections of Romantic Poetry.* Chapel Hill: U of North Carolina P, 1983. Print.

Franklin, Benjamin. *Meteorological Imaginations and Conjectures, in Memoirs of the Literary and Philosophical Society of Manchester.* London: T. Cadwell in the Strand, 1789. Print.

Garrard, Greg. *Ecocriticism.* New York: Routledge, 2004. Print.

Gaskell, Elizabeth. *North and South.* New York: Oxford UP, 1998. Print.

Gibson, William, and Bruce Sterling. *The Difference Engine.* New York: Bantam Books, 1991. Print.

Gigante, Denise. *Taste: A Literary History.* New Haven: Yale UP, 2005. Print.

Gilpin, William. "Three Essays on Picturesque Beauty, on Picturesque Travel, and on Sketching Landscape." *The Longman Anthology of British Literature.* Ed. S. Wolfson and P. Manning. Vol. 2A. New York: Longman, 2003. 506–10. Print.

Goellnicht, Donald C. *The Poet-Physician: Keats and Medical Science.* Pittsburgh: U of Pittsburgh P, 1983. Print.

Goethe, Johann W. *Scientific Studies.* Ed. and trans. Douglas Miller. Princeton: Princeton UP, 1987. Print.

Gould, Stephen Jay. *Time's Arrow, Time's Cycle: Myth and Metaphor in the Discovery of Geological Time.* Cambridge: Harvard UP, 1987. Print.

Gould, Stephen Jay, and Niles Eldredge. "Punctuated Equilibrium Comes of Age." *Nature* 366 (1993): 223–27. Print.

Goulding, Christopher. "A Volcano's Voice at Eton: Percy Shelley, James Lind, and Global Climatology." *Keats-Shelley Review* 17 (2003): 34–41. Print.

Grabo, Carl. *A Newton Among Poets: Shelley's Use of Science in Prometheus Unbound.* Chapel Hill: U of North Carolina P, 1930. Print.

Graham, Peter W. *Jane Austen and Charles Darwin: Naturalists and Novelists.* Burlington: Ashgate, 2008. Print.

Grattan, John, and Mark Brayshay. "An Amazing and Portentous Summer: Environmental and Social Responses in Britain to the 1783 Eruption of an Iceland Volcano." *Geographical Journal* 161.2 (1995): 125–34. Print.

Grove, Richard. *Green Imperialism: Colonial Expansion, Tropical Island Edens, and the Origins of Environmentalism, 1600–1860.* Cambridge: Cambridge UP, 1995. Print.

Halmi, Nicholas. "Mind as Microcosm." *European Romantic Review* 12.1 (2001): 43–52. Print.

Hansen, James. *Storms of My Grandchildren: The Truth About the Coming Climate Catastrophe and Our Last Chance to Save Humanity.* New York: Bloomsbury, 2009. Print.

Hansen, James, Makiko Sato, and Reto Ruedy. "Perceptions of Climate Change: The New Climate Dice." *Proceedings of the National Academy of Science* 119 (2012): 14726–727. Print.

Hardy, Thomas. *Tess of the d'Urbervilles.* New York: Modern Library, 1951. Print.

Harpp, Karen. "How Do Volcanoes Affect World Climate?" *Scientific American* 4 Oct. 2005. Web.

Harvey, William. *The Circulation of the Blood.* Trans. K. J. Franklin. London: Everyman's Library, 1963. Print.

Hawkes, Jacquetta. *A Land.* New York: Random House, 1951. Print.

Hayles, Katherine, ed. *Chaos and Order: Complex Dynamics in Literature and Science.* Chicago: U of Chicago P, 1991. Print.

Hector, Andy, and Rowan Hooper. "Ecology: Darwin and the First Ecological Experiment." *Science* 25 Jan. 2002: 639–40. Print.

Heringman, Noah. *Romantic Science: Literary Forms of Natural History*. Albany: SUNY P, 2003. Print.

Hooker, Jeremy. *Writers in a Landscape*. Cardiff: U of Wales P, 1996. Print.

Hutchinson, G. Evelyn. *The Ecological Theatre and the Evolutionary Play*. New Haven: Yale UP, 1965. Print.

Jefferies, Richard. *After London*. Charleston: Biblio-Bazaar, 2006. Print.

———. *At Home on the Earth: A New Selection of the Later Writings of Richard Jefferies*. Ed. Jeremy Hooker. Totnes: Green Books, 2001. Print.

———. *The Old House at Coate: And Other Hitherto Unpublished Essays*. Ed. Samuel J. Looker. Cambridge: Harvard UP, 1948. Print.

Johnson, James W. "Lyric." *The Princeton Encyclopedia of Poetry and Poetics, Enlarged Edition*. Ed. A. Preminger. Princeton: Princeton UP, 1974. 460–70. Print.

Keats, John. *Endymion: A Poetic Romance*. London: Taylor and Hessey, 1818. Print.

———. *Lamia, Isabella, The Eve of St. Agnes, and Other Poems*. London: Taylor and Hessey, 1820. Print.

———. *Selected Letters of John Keats*. Ed. Grant F. Scott. Cambridge: Harvard UP, 2002. Print.

Kerr, Richard. "Climate Change: Pushing the Scary Side of Global Warming." *Science* 8 (June 2007): 1412–15. Print.

Kerrigan, John. "Writing Numbers: Keats, Hopkins, and the History of Chance." *Keats and History*. Ed. N. Roe. Cambridge: Cambridge UP, 1995. 280–308. Print.

Kingsland, Sharon E. *Modeling Nature*. Chicago: U of Chicago P, 1995. Print.

Kingsolver, Barbara. *Prodigal Summer: A Novel*. New York: HarperCollins, 2000. Print.

Kricher, John. *The Balance of Nature: Ecology's Enduring Myth*. Princeton: Princeton UP, 2009. Print.

Lawler, Sharon P. "Ecology in a Bottle: Using Microcosms to Test Theory." *Experimental Ecology: Issues and Perspectives*. Ed. William J. Resetarits Jr. and Joseph Bernardo. New York: Oxford UP, 1998. 236–53. Print.

Leakey, Richard, and Roger Lewin. *The Sixth Extinction: Patterns of Life and the Future of Humankind*. New York: Doubleday, 1995. Print.

Levin, Susan M. *Dorothy Wordsworth and Romanticism*. New Brunswick: Rutgers UP, 1987. Print.

Levinson, Marjorie. *Keats's Life of Allegory: The Origins of a Style*. Oxford: Basil Blackwell, 1988. Print.

———. *The Romantic Fragment Poem: A Critique of Form*. Chapel Hill: U of North Carolina P, 1986. Print.

Little, Judy. *Keats as a Narrative Poet: A Test of Invention*. Lincoln: U of Nebraska P, 1975. Print.

Livingston, Ira. *Arrow of Chaos: Romanticism and Postmodernity*. Minneapolis: U of Minnesota P, 1997. Print.

Looker, Samuel J., and Crichton Porteous. *Richard Jefferies: Man of the Fields*. London: John Baker, 1965. Print.

Malthus, Thomas. *On the Principles of Population*. Vol. 1 of *The Works of Thomas Robert Malthus*. Ed. E. A. Wrigley and D. Souden. London: W. Pickering, 1986. Print.

Matthews, G. M. "A Volcano's Voice in Shelley." *ELH* 24.3 (1957): 191–228. Print.

McGann, Jerome. *The Beauty of Inflections: Literary Investigations in Historical Method and Theory*. Oxford: Clarendon, 1985. Print.

———. *The Romantic Ideology: A Critical Investigation*. Chicago: U of Chicago P, 1983. Print.

McGrath, Matt. "Ancient Irish Texts Show Volcanic Link to Cold Weather." *BBC News Online* 6 June 2013. Web.

McKusick, James C. *Green Writing: Romanticism and Ecology*. New York: St. Martin's, 2000. Print.

Merchant, Carolyn. *American Environmental History*. New York: Columbia UP, 2007. Print.

———. *The Death of Nature: Women, Ecology, and the Scientific Revolution*. San Francisco: Harper and Row, 1980. Print.

———. *Reinventing Eden: The Fate of Nature in Western Culture*. New York: Routledge, 2003. Print.

Miller, George, and Hugoe Matthews. *Richard Jefferies: A Bibliographic Study*. Brookfield: Ashgate, 1993. Print.

Milton, John. "Paradise Lost." *Complete Poems and Major Prose*. Ed. M. Hughes. Upper Saddle River: Prentice Hall, 1957. 173–470. Print.

Morgan, Mary S. "Afterword: Reflections on Exemplary Narratives, Cases, and Model Organisms." *Science Without Laws: Model Systems, Cases, Exemplary Narratives*. Ed. A. Creager, E. Lunbeck, and M. N. Wise. Durham: Duke UP, 2007. 264–74. Print.

Morris, William. *The Beauty of Life*. London: Brentham, 1974. Print.

Morton, Timothy. *The Ecological Thought*. Cambridge: Harvard UP, 2010. Print.

———. *Ecology Without Nature: Rethinking Environmental Aesthetics*. Cambridge: Harvard UP, 2007. Print.

Naeem, Saheed. "Gini in the Bottle." *Nature* 458 (2009): 579–80. Print.

NCDC (National Climatic Data Center). *Global Land and Ocean Annual Temperature Anomaly, 1880–2013*. Data file. Asheville: National Oceanic and Atmospheric Administration, U.S. Department of Commerce, 2012. Web.

Nutton, Vivian. "Medicine in the Greek World, 800–50 BC." *The Western Medical Tradition, 800 BC–1800 AD*. Ed. Lawrence I. Conrad et al. Cambridge: Cambridge UP, 2005. 11–38. Print.

Odum, Eugene, and Howard Odum. "Nature's Pulsing Paradigm." *Estuaries* 18.4 (1995): 547–55. Print.

Olson, Donald, Russell L. Doescher, and Marilynn S. Olson. "When the Sky Ran Red: The Story Behind 'The Scream.'" *Sky and Telescope* Feb. 2004: 28–35. Print.

O'Neill, Michael. "'When this warm scribe my hand': Writing and History in *Hyperion* and the *Fall of Hyperion*." *Keats and History*. Ed. N. Roe. Cambridge: Cambridge UP, 1995. 143–64. Print.

Oppenheimer, Clive. "Climatic, Environmental, and Human Consequences of the Largest Known Historic Eruption: Tambora Volcano (Indonesia), 1815." *Progress in Physical Geography* 27.2 (2003): 230–59. Print.

Ortony, Andrew. *Metaphor and Thought*. New York: Cambridge UP, 1979. Print.

Otis, Laura, ed. *Literature and Science in the Nineteenth Century: An Anthology*. Oxford: Oxford UP, 2002. Print.

Page, Norman. "The Ending of *After London*." *Notes and Queries* 32.3 (1985): 360–61. Print.

Paley, Morton. "Introduction." *The Last Man*. By Mary Shelley. Oxford: Oxford UP, 1998. Print.

Palmer, Trevor. *Perilous Planet Earth: Catastrophes and Catastrophism Through the Ages*. Cambridge: Cambridge UP, 2003. Print.

Palumbo, Donald E. *Chaos Theory, Asimov's Foundation and Robots, and Herbert's Dune: The Fractal Aesthetic of Epic Science Fiction*. Westport: Greenwood, 2002. Print.

Parker, Jo Alyson. *Narrative Form and Chaos Theory in Sterne, Proust, Woolf, and Faulkner*. Basingstoke: Palgrave Macmillan, 2007. Print.

Parrinder, Patrick. "From Mary Shelley to *The War of the Worlds*: The Thames Valley Catastrophe." *Anticipations: Essays on Early Science Fiction and Its Precursors*. Ed. D. Seed. Syracuse: Syracuse UP, 1995. 58–74. Print.

Pater, Walter. *The Renaissance: Studies in Art and Poetry*. Berkeley: U of California P, 1980. Print.

Perkins, David. *English Romantic Writers*. 2nd ed. Fort Worth: Harcourt Brace, 1995. Print.

Peterfreund, Stuart. "'Great Frosts and . . . Some Very Hot Summers': Strange Weather, the Last Letters, and the Last Days in Gilbert White's *Natural History of Selborne.*" *Romantic Science: The Literary Forms of Natural History.* Ed. N. Heringman. Albany: SUNY P, 2003. 85–108. Print.

Pollan, Michael. *The Omnivore's Dilemma: A Natural History of Four Meals.* New York: Penguin, 2006. Print.

Priestley, Joseph. *Experiments and Observations on Different Kinds of Air.* Birmingham: T. Pearson, 1790. Print.

Prigogine, Ilya, and Isabelle Stegners. *Order Out of Chaos: Man's New Dialogue with Nature.* New York: Bantam Books, 1984. Print.

"Review of *The Last Man* (1826)." *Monthly Review, or Literary Journal* ns 1 (Mar. 1826): 333–35. Reprinted in *Romantic Circles: Mary Wollstonecraft Shelley Chronology and Resource Site.* Ed. Neil Fraistat and Steven E. Jones. U of Maryland. Web.

Richardson, Alan. *British Romanticism and the Science of the Mind.* New York: Cambridge UP, 2001. Print.

———. "Keats and Romantic Science: Writing the Body." *The Cambridge Companion to Keats.* Ed. S. J. Wolfson. Cambridge: Cambridge UP, 2001. 230–45. Print.

Ricks, Christopher. *Keats and Embarrassment.* Oxford: Clarendon, 1974. Print.

Rosenberg, Charles E. *The Cholera Years: The United States in 1832, 1849, and 1866.* Chicago: U of Chicago P, 1987. Print.

Roszak, Betty, and Theodore Roszak. "Deep Form in Art and Nature." *The Green Studies Reader: From Romanticism to Ecocriticism.* Ed. L. Coupe. New York: Routledge, 2000. 223–27. Print.

Rothfield, Lawrence. *Vital Signs: Medical Realism in Nineteenth-Century Fiction.* Princeton: Princeton UP, 1992. Print.

Ruskin, John. "The Storm-Cloud of the Nineteenth Century." *Victorian Literature, 1830–1900.* Ed. D. Mermin and H. Tucker. Boston: Thompson and Heinle, 2002. 634–41. Print.

Ruston, Sharon. *Shelley and Vitality.* New York: Palgrave Macmillan, 2005. Print.

Sachs, Aaron. *The Humboldt Current: Nineteenth-Century Exploration and the Roots of American Environmentalism.* New York: Viking, 2006. Print.

Scott, Heidi C. M. "Ecological Microcosms in Richard II." *Explicator* 67.4 (2009): 267–71. Print.

Self, Stephen, and Michael Rampino. "The 1883 Eruption of Krakatau." *Nature* 294 (1981): 699–704. Print.

Shakespeare, William. *Hamlet.* Folger Shakespeare Library. New York: Simon and Schuster, 2003. Print.

———. *Macbeth.* Folger Shakespeare Library. New York: Simon and Schuster, 2004. Print.

———. *Richard II.* Folger Shakespeare Library. New York: Simon and Schuster, 2005. Print.

Shelley, Mary W. *Frankenstein.* Oxford: Oxford UP, 1998. Print.

———. *The Last Man.* Oxford: Oxford UP, 1998. Print.

Shelley, Percy Bysshe. *The Complete Poetry of Percy Bysshe Shelley.* Ed. D. Reiman and N. Fraistat. Baltimore: Johns Hopkins UP, 2000. Print.

———. *Prometheus Unbound: A Lyrical Drama in Four Acts, with Other Poems.* London: C. and J. Ollier, 1820. Print.

Simberloff, D., and E. O. Wilson. "Experimental Zoogeography of Islands—Colonization of Empty Islands." *Ecology* 50 (1969): 278–96. Print.

Sinclair, George. *Hortus gramineus woburnensis.* 2nd ed. London: James Ridgeway, 1824. Print.

Smith, Robert Angus. *Air and Rain: The Beginnings of a Chemical Climatology.* London: Longman, 1872. Print.

Snyder, Gary. *A Place in Space: Ethics, Aesthetics, and Watersheds.* New York: Counterpoint, 1995. Print.

Soden, Brian J., and Isaac M. Held. "An Assessment of Climate Feedbacks in Coupled Ocean-Atmosphere Models." *Journal of Climate* 19 (2006): 3354–60. Print.

Soper, Kate. *What Is Nature?* Cambridge: Blackwell, 1995. Print.

Spencer, Herbert. *Education: Intellectual, Moral, and Physical.* Totowa: Littlefield and Adams, 1965. Print.

Srivastava, Diane S., et al. "Are Natural Microcosms Useful Model Systems for Ecology?" *Trends in Ecology and Evolution* 19.7 (2004): 379–84. Print.

Sydney Morning Herald. "Qantas Cancels Flights for a Third Day." 18 Apr. 2010. Web.

Tansley, A. G. "The Use and Abuse of Vegetational Concepts and Terms." *Ecology* 16.3 (1935): 284–307. Print.

Thomas, Sophie. "The Ends of the Fragment, the Problem of the Preface: Proliferation and Finality in *The Last Man*." *Mary Shelley's Fictions: From Frankenstein to Falkner.* Ed. M. Eberle-Sinatra. New York: St. Martin's, 2000. 22–38. Print.

Vestheim, Hege, et al. "Lack of Response in a Marine Pelagic Community to Short-term Oil and Contaminant Exposure." *Journal of Experimental Marine Biology and Ecology* 416–17 (2012): 110–14. Print.

Walker, Dan. "When a Killer Cloud Hit Britain." *BBC News* 19 Jan. 2007. Web.

Ward, Nathaniel. *On the Growth of Plants in Closely Glazed Cases.* London: J. Van Voorst, 1852. Print.

Wearden, Graeme. "Ash Cloud Costing Airlines £130m a Day." *Guardian* 16 Apr. 2010. Web.

Weber, A. S., ed. *Nineteenth-Century Science: A Selection of Original Texts.* Orchard Park: Broadview, 2000. Print.

Weisman, Alan. *The World Without Us.* New York: St. Martin's, 2007. Print.

Wells, H. G. *The Time Machine.* New York: Penguin, 2005. Print.

Wells, H. G., Julian S. Huxley, and G. P. Wells. *The Science of Life.* Garden City: Doubleday, Doran, 1931. Print.

White, Gilbert. *The Illustrated Natural History and Antiquities of Selborne.* New York: Penguin Classics, 1987. Print.

White, Hayden. *Metahistory: The Historical Imagination in Nineteenth-Century Europe.* Baltimore: Johns Hopkins UP, 1973. Print.

Williams, Raymond. *The Country and the City.* New York: Oxford UP, 1975. Print.

Wilson, Edward O. *Consilience: The Unity of Knowledge.* New York: Knopf, 1998. Print.

Winchester, Simon. *Krakatoa: The Day the World Exploded; August 27, 1883.* New York: Harper, 2003. Print.

Wittebolle, Lieven, et al. "Initial Community Evenness Favours Functionality Under Selective Stress." *Nature* 458 (2009): 623–26. Print.

Wordsworth, William. *The Complete Poetical Works of William Wordsworth, 1806–1815.* Ed. Andrew J. George. Vol. 5. Boston: Houghton Mifflin, 1919. Print.

———. *Home at Grasmere.* Ed. B. Darlington. Ithaca: Cornell UP, 1977. Print.

———. "Lines Written a Few Miles Above Tintern Abbey, on Revisiting the Banks of the Wye During a Tour, July 13, 1798." *Lyrical Ballads, with a Few Other Poems.* By William Wordsworth and Samuel T. Coleridge. London: J. & A. Arch, 1798. 201. Print.

———. *The Prelude, or Growth of a Poet's Mind.* 2nd ed. Ed. E. de Selincourt and H. Darbishire. London: Oxford UP, 1959. Print.

Worster, Donald. *Nature's Economy: A History of Ecological Ideas.* New York: Cambridge UP, 1994. Print.

Zellmer, A. J., T. F. H. Allen, and K. Kesseboehmer. "The Nature of Ecological Complexity: A Protocol for Building the Narrative." *Ecological Complexity* 3 (2006): 171–82. Print.

Žižek, Slavoj. *The Parallax View.* Cambridge: MIT P, 2006. Print.